Nuclear Development

Carbon Pricing, Power Markets and the Competitiveness of Nuclear Power

© OECD 2011
NEA No. 6982

NUCLEAR ENERGY AGENCY
ORGANISATION FOR ECONOMIC CO-OPERATION AND DEVELOPMENT

ORGANISATION FOR ECONOMIC CO-OPERATION AND DEVELOPMENT

The OECD is a unique forum where the governments of 34 democracies work together to address the economic, social and environmental challenges of globalisation. The OECD is also at the forefront of efforts to understand and to help governments respond to new developments and concerns, such as corporate governance, the information economy and the challenges of an ageing population. The Organisation provides a setting where governments can compare policy experiences, seek answers to common problems, identify good practice and work to co-ordinate domestic and international policies.

The OECD member countries are: Australia, Austria, Belgium, Canada, Chile, the Czech Republic, Denmark, Estonia, Finland, France, Germany, Greece, Hungary, Iceland, Ireland, Israel, Italy, Japan, Luxembourg, Mexico, the Netherlands, New Zealand, Norway, Poland, Portugal, the Republic of Korea, the Slovak Republic, Slovenia, Spain, Sweden, Switzerland, Turkey, the United Kingdom and the United States. The European Commission takes part in the work of the OECD.

OECD Publishing disseminates widely the results of the Organisation's statistics gathering and research on economic, social and environmental issues, as well as the conventions, guidelines and standards agreed by its members.

> *This work is published on the responsibility of the OECD Secretary-General.*
> *The opinions expressed and arguments employed herein do not necessarily*
> *reflect the views of all member countries.*

NUCLEAR ENERGY AGENCY

The OECD Nuclear Energy Agency (NEA) was established on 1 February 1958. Current NEA membership consists of 30 OECD member countries: Australia, Austria, Belgium, Canada, the Czech Republic, Denmark, Finland, France, Germany, Greece, Hungary, Iceland, Ireland, Italy, Japan, Luxembourg, Mexico, the Netherlands, Norway, Poland, Portugal, the Republic of Korea, the Slovak Republic, Slovenia, Spain, Sweden, Switzerland, Turkey, the United Kingdom and the United States. The European Commission also takes part in the work of the Agency.

The mission of the NEA is:

– to assist its member countries in maintaining and further developing, through international co-operation, the scientific, technological and legal bases required for a safe, environmentally friendly and economical use of nuclear energy for peaceful purposes, as well as

– to provide authoritative assessments and to forge common understandings on key issues, as input to government decisions on nuclear energy policy and to broader OECD policy analyses in areas such as energy and sustainable development.

Specific areas of competence of the NEA include the safety and regulation of nuclear activities, radioactive waste management, radiological protection, nuclear science, economic and technical analyses of the nuclear fuel cycle, nuclear law and liability, and public information.

The NEA Data Bank provides nuclear data and computer program services for participating countries. In these and related tasks, the NEA works in close collaboration with the International Atomic Energy Agency in Vienna, with which it has a Co-operation Agreement, as well as with other international organisations in the nuclear field.

Foreword

As part of the global effort to limit CO_2 emissions that prompt climate change, a key objective of carbon pricing is to decarbonise electricity generation and to make investments in low-carbon power sources more attractive. In OECD countries, such investment is increasingly being financed by private investors in markets with liberalised electricity prices. An earlier IEA/NEA study, *Projected Costs of Generating Electricity: 2010 Edition*, had already demonstrated the competiveness of nuclear power under the assumption of a carbon price of USD 30/tCO_2 with the help of the levelised cost methodology, which reflects the conditions of regulated markets.

This new NEA study, prepared under the oversight of the Working Party on Nuclear Energy Economics (WPNE), instead asks the question of "what is the most profitable technology for baseload power generation from the point of view of a private investor in the context of liberalised markets with volatile electricity prices and carbon pricing in place?" It analyses this question both under the assumption of a carbon market with volatile prices for CO_2 permits as well as under the assumption of a stable carbon tax. This is the first carbon pricing study using real market data, as it benefits from access to daily price data on European markets for electricity, gas, coal and carbon during a period stretching from July 2005 to May 2010. This encompasses very nearly the first five years of the European Emissions Trading System (EU ETS), the world's foremost carbon trading framework.

The results of the study converge on one major finding: even with modest carbon pricing, future competition in power generation will take place between nuclear energy and gas-fired power generation, with standard coal-fired power plants no longer being profitable. The outcome of the nuclear versus gas competition hinges, in addition to carbon pricing, on a number of factors which include the overnight costs for nuclear power plant construction, financing costs, gas prices, profit margins in the electricity sector due to monopoly power, the price of electricity or the likelihood of a pervasive deployment of carbon capture and storage (CCS). One can summarise these considerations in the following manner. *Nuclear energy is competitive with natural gas for baseload power generation, as soon as one of the three main parameters – investment costs, prices or CCS – acts in its favour. It will dominate the competition as soon as two out of three categories act in its favour.*

It is important to recall that, according to the parameters of this study, a new nuclear power plant being commissioned in 2015 would produce electricity until 2075. During that period it is likely that gas prices will be higher than today and that coal-fired power plants will be equipped with carbon capture and storage. Readers are thus invited to pay particular attention to the CCS analysis in the second part of Chapter 7.

For policy makers the study provides a number of counter-intuitive but robust insights that should be heeded to improve the long-term efficiency of policy making in the power sector when markets are liberalised:

1. At current gas prices and in the absence of carbon capture and storage for coal-fired power plants, carbon pricing is most effective in enhancing the competitiveness of nuclear energy in a range of EUR 30-50 (USD 43-72) per tonne of CO_2.

2. Strong competition in electricity markets leading to low mark-ups above variable costs enhances the competitiveness of nuclear power.

3. Pervasive deployment of carbon capture and storage (CCS) substantially *improves* the competitiveness of nuclear power as it decreases the margins of gas-fired power generation.

Last but not least, one needs to underline the importance of electricity price stability. Due to the cost structure of nuclear power, risk-averse investors may opt for fossil-fuel-fired power generation instead of nuclear, *even in cases where nuclear energy would be the least-cost option* (according to levelised cost methodology). Liberalised electricity markets with uncertain prices can lead to different decisions being taken by risk-averse private investors than by governments with a longer-term view. This especially concerns investments in low-carbon technologies with high fixed costs. And while only electricity market liberalisation can provide the dynamism and competitive pressure for markets to radically change the structure of power supplies in the next two decades, policy makers should consider means such as long-term contracts, price guarantees or customer finance in order to let capital-intensive, low-carbon technologies such as nuclear and certain renewable energies compete on an equal footing with less capital-intensive, fossil-fuel technologies.

Overall the study provides an array of results under a series of different assumptions and configurations related to the main parameters mentioned earlier, all based on empirical market data. Other reasonable assumptions and configurations can certainly be conceived but the choices in this study seem reasonable and justifiable. The ultimate role of this study is thus to provide a template for the further study of the economic conditions for a transition towards low-carbon electricity sectors in OECD/NEA countries.

Acknowledgements

This study was written by Dr. Jan Horst Keppler, Principal Economist, and Dr. Claudio Marcantonini, Nuclear Energy Expert, at the OECD Nuclear Energy Agency (NEA). Dr. Ron Cameron, Head of the NEA Nuclear Development Division, provided managerial oversight as well as substantial comment throughout the process.

The study was part of the Programme of Work of the Committee for Technical and Economic Studies on Nuclear Energy Development and the Fuel Cycle (NDC). It was supervised by the NEA Working Party on Nuclear Energy Economics (WPNE), which is a subcommittee of the NDC, under its Chairmen Mr. Matthew Crozat and Professor Alfred Voss. Throughout the study, the WPNE ensured the consistency of the study's messages with those of previous NEA publications such as *Projected Costs of Generating Electricity* (2010, together with IEA). The document was endorsed for publication by the NDC.

The authors would like to thank Sovann Khou, Powernext, Charlotte de Lorgeril, Sia Conseil and Susann Zimmer, EEX, for their kindness in providing the raw data sets for the establishment of consistent daily price data over five years in four different markets. Without their help, the study would not have been feasible.

The authors would further like to thank the participants of the international workshop on "Carbon Pricing, Power Markets and the Competitiveness of Nuclear Power: Strengths and Weaknesses under Different Price Scenarios" which was held in Paris on 11 January 2011 and helped to clarify and strengthen several aspects of the study. Special thanks go to the presenters at the workshop: Professor Richard J. Green, University of Birmingham, Professor David M. Newbery, University of Cambridge, Professor John Parsons, Massachusetts Institute of Technology (MIT), and Dr. Fabien Roques, IHS CERA.

Table of contents

Executive summary

The pricing of greenhouse gas emissions has increasingly become a reality in industrialised countries trying to attain their emission reduction targets defined under the 1997 Kyoto Protocol. Given that carbon dioxide (CO_2) emissions, also referred to as "carbon emissions", constitute the largest and most easily measurable share of greenhouse gas emissions (76% of the global total), it is no surprise that emission reduction efforts are concentrated in this area. Roughly 80% of CO_2 emissions are due to the burning of fossil fuels and of these roughly 40% are due to the generation of electricity and heat in the power sector, where the burning of coal contributes about three quarters of all carbon emissions. The question is what will be the role of nuclear energy once efforts to reduce these emissions have begun in earnest.

The accident at the Fukushima Daiichi nuclear power plant in Japan in March 2011 has of course questioned a number of assumptions in the nuclear power industry and in the energy industry at large. Nevertheless, the reality of climate change and of measures to reduce greenhouse gas emissions, among which carbon pricing is the most prominent and likely to be the most efficient, will not go away (see Box ES.1). In addition, the powerful trend in OECD countries towards more liberalised power markets that is driven by long-term developments in information technology, network management, regulatory and managerial progress, and increased consumer awareness will continue.

Box ES.1: How realistic is the NEA's carbon price analysis after Fukushima?

This NEA study works with a first-of-a-kind (FOAK) case and an industrial maturity case for Generation III+ reactors which can be interpreted as the upper and lower bounds of the future investment costs for nuclear energy. The precise cost of future reactors will be difficult to determine for some time for two reasons. Firstly, deployment of the new Generation III and III+ reactors will generate economies of scale, but how much precisely is difficult to say. Secondly, the partial fuel meltdown at three nuclear plants after the failure of the cooling systems in the wake of a major earthquake and a very large tsunami at the Fukushima Daiichi nuclear power plant in Japan will trigger a regulatory review of the safety features that will be required for existing as well as new nuclear power plants. It is too soon to draw conclusions on the cost implications of the requirements emanating from the lessons learnt at Fukushima. While there might be some impact in terms of added costs, there is reason to think that it might be limited given that Generation III+ reactors already have a number of safety features such as multiple (up to four) independent cooling systems, cooling systems that work by natural convection (passive cooling), core catchers and strong outer containment domes (in addition to the interior reactor containment vessel) able to withstand high pressures. In other words, the assumptions of this study would seem to remain a valid range for new European nuclear reactors in the coming years.

The basic question of this study, "what will be the impact of carbon pricing on the competitiveness of nuclear energy compared to coal- and gas-fired power generation in a context of liberalised electricity markets?" is thus as valid as ever. This study, which was started in September 2010 under the oversight of the NEA Working Party on Nuclear Energy Economics, is also the first-ever attempt to tackle the question of the competitiveness of different power generation technologies under carbon pricing on the basis of empirical data. In doing so, it analyses daily data from European power and carbon markets during a period stretching from July 2005 to May 2010. This encompasses very nearly the first five years of the European Emissions Trading System (EU ETS), the world's foremost carbon trading framework (see Figure ES.1). Nevertheless, many of the conclusions are applicable to other

OECD regions to the extent that power market liberalisation has taken hold. The study also provides calculations of the levelised cost of electricity (LCOE) for all three OECD regions, which constitute an important benchmark for cost competitiveness in regulated power markets.

Figure ES.1: European prices for electricity, carbon, gas and coal
2005-10

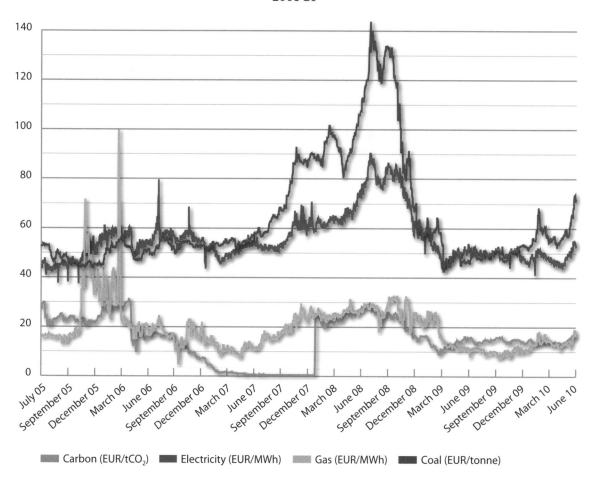

■ Carbon (EUR/tCO$_2$)   ■ Electricity (EUR/MWh)   ■ Gas (EUR/MWh)   ■ Coal (EUR/tonne)

This NEA assessment of the competitiveness of nuclear energy against coal- and gas-fired generation under carbon pricing consistently adopts the viewpoint of a private investor seeking to maximise the return of his/her invested funds. The study broadly confirms, albeit in far greater detail and considering a much greater number of variables, the results of the *Projected Costs of Generating Electricity* (IEA/NEA, 2010). And while the Projected Costs study adopted a concept of social resource cost based on the LCOE methodology rather than on private profit maximisation, one basic conclusion remains the same: competition in electricity markets is today being played out between nuclear energy and gas-fired power generation, with coal-fired power generation not being competitive once carbon pricing is introduced (see Figure ES.2). Whether nuclear energy or natural gas comes out ahead in this competition depends on a number of assumptions, which even for variations inside entirely reasonable ranges, can yield very different outcomes.

Figure ES.2: Carbon pricing and the competitiveness of nuclear energy in OECD Europe
LCOE of different power generation technologies at a 7% discount rate

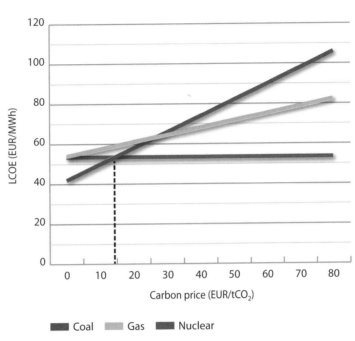

Source: Adapted from IEA/NEA, 2010.

In order to assess the profitability of different options for power generation, the study employs three gradually more complete methodologies beyond the LCOE approach: a profit analysis looking at historic returns over the past five years, an investment analysis projecting the conditions of the past five years over the lifetime of plants and a carbon tax analysis (differentiating the investment analysis for different carbon prices) looking at the issue of competitiveness from different angles. They show that the competitiveness of nuclear energy depends on a number of variables which in different configurations determine whether electricity produced from nuclear power or from combined-cycle gas turbines (CCGTs) generates higher profits for its investors. They are:

1. *Overnight costs:* the profitability of nuclear energy as the most capital-intensive of the three technologies depends heavily on its overnight costs.[1] This is a characteristic that it shares with other low-carbon technologies such as renewable energies, but the latter are not included in this comparison. The study reflects the importance of capital costs by working with a FOAK case and an industrial maturity case, where the latter's capital cost is two-thirds of the former's.

2. *Financing costs:* since the Projected Costs study nothing has changed on this point. Financing costs have a very large influence on the costs and profitability of nuclear energy. Nevertheless, the study does not concentrate on this well-known point but works (except for one illustrative case) with a standard capital cost of 7% real throughout the study.

---

1. Capital costs are a function of overnight costs (which include pre-construction or owner's cost, engineering, procurement and construction costs as well as contingency costs) and interest during construction (IDC). The latter depends, of course, on financing costs as discussed under the next point.

3. *Gas prices:* what capital costs are to the competitiveness of nuclear energy, gas prices are to the competitiveness of gas-fired power generation, which spends a full two-thirds of its lifetime costs on fuel. If gas prices are low, gas-fired power generation is very competitive indeed. If they are high, nuclear energy is far ahead. The study reflects this fact by working with a low gas price case and a high gas price case in addition to the base case scenario.

4. *Carbon prices:* low and medium-high carbon prices, up to EUR 50 per tonne of CO_2 (tCO_2) increase the competitiveness of nuclear power. However, in contrast to the conclusions of the LCOE methodology employed in the Projected Costs study, high carbon prices do not unequivocally improve the competitiveness of nuclear power in a market environment. As carbon pricing makes coal with its high carbon content the marginal fuel, the revenues of gas increase faster than its cost, with an overall increase in profitability that matches that of nuclear and can surpass it at very high carbon prices.

5. *Profit margins* or "mark-ups" are the difference between the variable costs of the marginal fuel and the electricity price, and are a well-known feature of liberalised electricity markets. They have a very strong influence on the competitiveness of the marginal fuel, either gas or coal, for which they single-handedly determine profits. The level of future profit margins can thus determine the competitiveness between nuclear energy and gas.

6. *Electricity prices:* in a liberalised electricity market, prices are a function of the costs of fossil fuels (natural gas and coal), carbon prices and mark-ups. The higher they are, the better nuclear energy fares, both absolutely and relatively. This is also due to the fact that higher electricity prices go along with higher prices for fossil fuels and carbon.

7. *Carbon capture and storage (CCS):* the standard investment and carbon tax analyses do not assume the existence of pervasive CCS for coal-fired power plants. However, an alternative scenario does and it shows that CCS will remarkably strengthen the relative competitiveness of nuclear energy against gas-fired power generation. The profitability of gas declines significantly once it substitutes for coal as the marginal fuel at high carbon prices.

The particular configuration of these seven variables will determine the competitive advantage of the different power generation options. The profit analysis showed that during the past five years, nuclear energy has made very substantive profits due to carbon pricing (see Figure ES.3). These profits are far higher than those of coal and gas, even though the latter did not have to pay for their carbon emission permits during the past five years of Phase I and Phase II of the EU ETS. Operating an existing nuclear power plant in Europe today is very profitable.

The conclusion that an existing nuclear power plant is highly profitable under carbon pricing is independent of the particular carbon pricing regime both in absolute and in relative terms. Given that nuclear power would not have to acquire carbon permits under any regime, its profits would not change as long as electricity prices stay the same. Profits would change instead for coal- and gas-fired generation. The switch to auctioning permits under the EU ETS in 2013, which will oblige emitters actually to pay for their emissions, will thus increase the competitive advantage of nuclear energy due to carbon pricing. Substituting an emissions trading scheme characterised by volatile prices with a stable carbon tax equivalent to the average trading price would actually increase the volatility of profits for coal and gas and thus increase the relative competitiveness of nuclear energy even further. Contrary to the opinion that nuclear would be better served by a stable tax, the empirical evidence indicates that nuclear energy does at least as well under carbon trading, including when carbon prices are volatile.

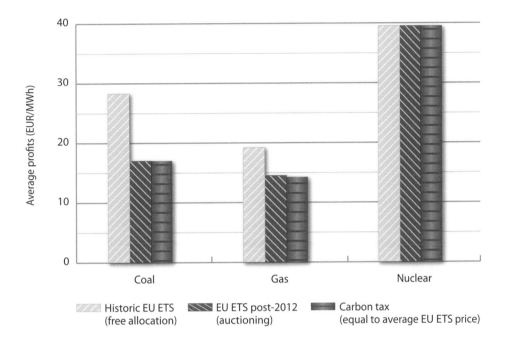

Figure ES.3: Average profits with suspension option

However, the profit analysis does not consider investment costs. It is more difficult to summarise the results for the investment and the carbon tax analysis, which both take into account the investment costs and compute the costs and benefits over the lifetime of the different plants. Again, a new coal plant is highly unlikely to be a competitive or even a profitable technology option under the price conditions prevailing during the 2005-10 period once it has to pay for its carbon emissions. Concerning the competition between nuclear energy and gas-fired power generation measured in terms of an appropriately defined profitability index (PI), one needs to differentiate and to specify the particular configuration of the seven variables presented above. If the seven variables above are grouped in three broad categories, investment costs, electricity prices as a function of gas, and carbon prices and CCS – then one may summarise the results of this study in the following manner. *Nuclear energy is competitive with natural gas for baseload power generation as soon as one of the three categories – investment costs, prices or CCS – acts in its favour. It will dominate the competition as soon as two out of three categories act in its favour.*

It is important to recall that according to the parameters of this study, a new nuclear power plant being commissioned in 2015 would produce electricity until 2075. While final appreciations are the prerogative of each individual investor, there is clearly a very strong probability that gas prices will be considerably higher than today and that coal-fired power plants will be consistently equipped with carbon capture and storage during that period. Readers are thus invited to pay particular attention to the CCS analysis in the second part of Chapter 7.

The competition between nuclear energy and gas-fired power generation remains characterised by the dependence of each technology's profitability on different scenarios. Gas, which is frequently the marginal fuel, makes modest profits in many different scenarios, which limits downside as well as upside risk. The small size of its fixed costs does not oblige it to generate very large profit margins.

High electricity prices are not necessarily a source of significant additional profits as they frequently result precisely from high gas prices. Nuclear energy is in the opposite situation, where its profitability depends almost exclusively on electricity prices. Its high fixed costs and low and stable marginal costs mean that its profitability rises and falls with electricity prices (see Figure ES.4).

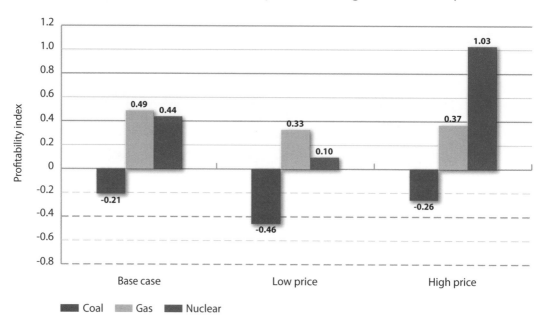

Figure ES.4: Profitability index in different electricity price scenarios
7% real discount rate, industrial maturity case and average 2005-10 carbon price

Carbon pricing will, of course, increase the competitiveness of nuclear energy against coal and to a lesser extent against gas. In the competition between nuclear energy and gas, carbon pricing will favour nuclear, in particular in a range up to EUR 50 per tonne of CO_2 (in comparison, the five-year average on the EU ETS was slightly over EUR 14). Beyond that range, coal-fired power generation will consistently set electricity prices and gas-fired power plants will thus earn additional rents faster than their own carbon costs increase. This may, at very high carbon prices, enable gas to even surpass nuclear energy (see Figure ES.5). While coherent at the level of the modelling exercise, it should be said that market behaviour and cost conditions at carbon prices above EUR 50 per tonne of CO_2 are quite uncertain, and results for any configuration in that range should be considered with caution. One would, for instance, expect that high carbon prices applied consistently over time would generate a number of dynamic effects and technological changes, such as a quicker penetration of carbon capture and storage (CCS). This would substantially alter results by enhancing the relative competitiveness of nuclear against gas (see Figure ES.6).

Figure ES.5: Evolution of profitability indices in the base case scenario
Constant profit margin of EUR 10, 7% real discount rate and industrial maturity case

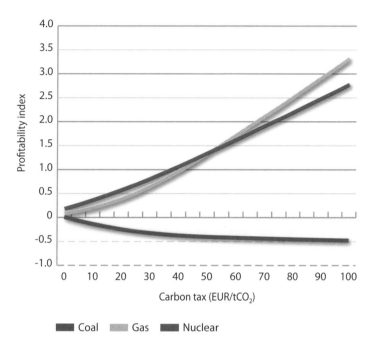

Figure ES.6: Evolution of profitability indices in the CCS base case scenario
Constant profit margin of EUR 10, 7% real discount rate, industrial maturity case and coal with carbon capture

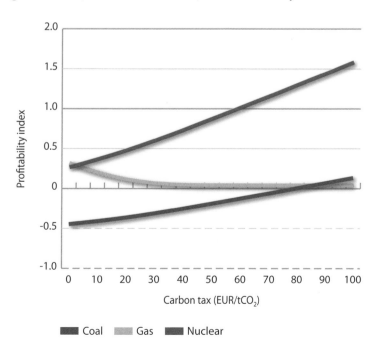

For investors, it is thus important to make their own assessment of the probability of different capital costs and price scenarios. If nuclear succeeds in limiting overnight costs and electricity prices in Europe stay high, nuclear energy is by far the most competitive option. With high overnight costs and low electricity prices, only a strong logic of portfolio diversification could motivate an argument in its favour. As far as prices are concerned, it is quite likely that European electricity prices will stay high or even increase in the foreseeable future. The progressive exit from both fossil fuels and nuclear energy in Germany, Europe's biggest market, will inevitably push electricity prices higher, which in conjunction with carbon pricing opens opportunities for nuclear energy in other European countries. Similar dynamics may also assert themselves in the United States, where ambitious greenhouse gas reduction targets also ensure a floor under electricity prices.

A high electricity price scenario is thus likely, but by no means assured. In this context, policy makers need to be aware of the fact that the profitability of nuclear energy in liberalised electricity markets depends on specific electricity price scenarios. It is thus not unthinkable that risk-averse private investors may opt for fossil-fuel-fired power generation instead of nuclear, *even in cases where nuclear energy would be the least-cost option over the lifetime of the plant.* Liberalised electricity markets with uncertain prices can lead to different decisions being taken by risk-averse private investors than by governments with a longer-term view. Care has to be taken to reflect the specificities of high fixed cost, low-carbon technologies such as nuclear energy and certain renewables in the process through appropriate measures, for example, long-term contracts for electricity provision. Otherwise, the risk of private and social optimality disconnecting is very real.

An additional aspect of public policy making concerns the profit margins or mark-ups of electricity prices over the variable costs of the marginal fuel which benefit, in particular, the competitiveness of the last fuel in the merit order. Regardless of whether they are an expression of spontaneous or consciously constructed monopoly power, nuclear energy is favoured by limiting these welfare-reducing mark-ups. Market opening and competition in the provision of baseload power favour the competitiveness of nuclear energy.

In the end, the outcome of the competition between nuclear energy and gas-fired power generation (coal-fired power generation being uncompetitive under carbon pricing) depends on a number of key parameters such as investment costs and prices. The profitability of either nuclear energy or gas-fired power generation, however, cannot be assessed independently of the scenario in which they are situated. Given the realities of the large, integrated utilities that dominate the European power market, which need to plan ahead for a broad range of contingencies, the implications are straightforward. Risk minimisation implies that utilities need to diversify their generation sources and to adopt a portfolio approach. Any utility would thus be advantaged by a portfolio approach. Such diversification would not only limit financial investor risk, but also a number of non-financial risks (climate change, security of supply, accidents). Hence, portfolio approaches and the integration of non-financial risks will both be important topics for future research at the NEA and in the wider energy community.

Chapter 1

Introduction

This NEA study assesses the competitiveness of nuclear energy in liberalised power markets with carbon pricing. It is in many ways a follow-up study to the joint IEA/NEA study, *Projected Costs of Generating Electricity: 2010 Edition* that established the costs of different power generation technologies by comparing the levelised costs of generating electricity. There exist, however, two crucial differences with the Projected Costs study. First, this study concentrates on the profit calculations of a private investor in an environment with volatile prices rather than on the levelised costs of electricity. Second, this analysis is based on empirical price and cost data from deregulated European electricity markets during the 2005-10 period rather than on the contribution by member country governments.

Private cost-benefit calculations are performed both for existing plants ("profitability analysis") and for yet to built, new plants ("investment analysis") assuming the historical cost and price conditions of the past five years. The "carbon tax analysis" will perform cost-benefit calculations also for carbon prices other than the average carbon price of EUR 14 that prevailed from 2005 to 2010. In all three areas, the study aims as much as possible to work with data from actual electricity and carbon markets.[1] While the stated objective of assessing the competitiveness of nuclear energy from the perspective of a private investor seems straightforward enough, in practice, a number of issues need to be clarified in advance. This includes the nature of the carbon pricing scheme (a trading system or a tax), the metric for profitability (net present value, internal rate of return, profitability index, etc.) and, of course, assumptions about investment and variable costs.

Assessing the competitiveness of different power technologies under carbon pricing also depends on the particular mechanism chosen to integrate the social cost of climate change inducing CO_2 emissions into the decisions of utilities and investors. The impact of a stable carbon tax is not necessarily the same as that of an emissions trading system such as the EU ETS with volatile prices. Chapter 5 will show that the effect, while not overwhelming, is nevertheless significant and merits being taken into account in policy-making decisions.

---

1.   Market prices for electricity, CO_2, coal and gas have been taken from European wholesale markets during the period 2005-10. Data on investment and operations and maintenance (O&M) costs are from the 2010 IEA/NEA study *Projected Costs of Generating Electricity* (see Chapter 3 for more details).

1.1 Social resource costs versus private profitability calculations in a real market environment

In principle, there is widespread agreement that carbon pricing can reshape electricity sectors. The notion is very intuitive that higher costs for carbon emissions, whether in form of a tax or the price of a quota, increase the production costs of carbon-intensive producers such as coal and gas (oil only produces 4% of electricity in OECD countries) and enhance the competitiveness of low-carbon producers such as nuclear and renewable energies. In the following, this study concentrates on comparing the costs and profitability of nuclear power with those of coal- and gas-based power. While the importance of carbon pricing is widely acknowledged, there has been to date little empirical work on the issue, frequently due to the lack of coherent sets of multi-year market data for fuel, carbon and electricity markets. Benefiting from the access to precisely this kind of data, this study is thus able to provide the first systematic empirically based analysis of the competitiveness of nuclear power in liberalised electricity markets with carbon pricing.

In terms of assessing the competitiveness of different technologies under carbon pricing, the IEA/NEA study *Projected Costs of Generating Electricity* was a first important step as it provided data on the LCOE for a large number of different power generation technologies in OECD countries. The Projected Costs study had already assumed a price of USD 30 per tonne of CO_2. On this basis, the NEA Secretariat performed a number of sensitivity analyses with different carbon prices, which are presented in Chapter 4 of the present study.

However, the investment decisions that would follow from the notion of costs used in the Projected Costs study are different from those a private investor would make on the basis of his/her own profitability calculations. Assessing the impact of carbon pricing on the competitiveness of different power technologies in *Projected Costs of Generating Electricity* provides, of course, important insights in its own right. It needs to be understood, however, that the carbon cost sensitivity analyses are based on a specific notion of costs referred to as "social resource cost". By definition, the LCOE derived in the Projected Costs study indicates the price per unit of electricity that would allow a specific power generating investment to break even if this price would be paid for output during the lifetime of the project. LCOE calculations thus provide an indication to policy makers and modellers of the real resources that are required for a given investment under the assumption of stable electricity prices in electricity systems with rate-of-return regulation.

LCOE calculations undoubtedly provide important information for framing long-term policy choices and *Projected Costs of Generating Electricity* is rightly a widely used input for policy discussions and long-term energy system modelling. LCOE, however, is a very imperfect indicator for the choices a private investor needs to make on the basis of the likely profitability of different technologies in a liberalised electricity market. Instead, LCOE is a good indicator for investment choices in a regulated electricity market with stable and predictable electricity prices.[2]

The difference between the costs generated by the LCOE methodology and private investment costs consists of two essential issues: i) price risk in liberalised electricity markets and ii) the specific price formation mechanisms in liberalised electricity markets. Price risk is a crucial difference

---

2. This holds as long as technologies are compared for similar uses at equivalent utilisation rates. The variability of demand over the day and the year coupled with the non-storability of electricity mean that electricity production is subdivided into constant baseload production and intermittent peakload production. This means one technology can have the lowest LCOE for a high utilisation rate (baseload) and another one may have the lowest LCOE for a low utilisation rate (peakload). *Projected Costs of Generating Electricity* compares nuclear-, coal- and gas-based production on the basis of the common assumption of an utilisation rate of 85% (baseload). In such a case, LCOE is indeed a valid indicator for regulators interested in choosing the technology which minimises the social resource cost.

between LCOE calculations and private cost calculations. Given that by definition the LCOE is equivalent to the constant price that would allow an investor to break even, there is no price risk involved. In liberalised markets where prices are volatile, investors are confronting a different situation. Even when computed future profits in function of uncertain future prices are overall slightly positive, there is still a non-negligible risk that the final outcome would show a loss. For a risk-averse investor, however, the probability of bankruptcy needs to be minimised as much as possible. This means he/she will include into his/her choice of technology not only how price volatility impacts the average profitability of his/her investment but also the range of the different outcomes and, in particular, the risk of bankruptcy. Other things being equal (for instance LCOE), a higher ratio of fixed to variable costs increases price risk as technologies with higher variable costs have the possibility to evade price risk simply by stopping production when prices are low.[3]

1.2 Special issues in electricity markets

Price formation in liberalised electricity markets is significantly impacted by two particularities that distinguish electricity markets from most other markets. First, electricity is a non-storable good which creates high price volatility in the segment of the market that allocates production at short-notice, the spot or day-ahead market. Second, the variable or marginal costs of different technologies vary widely. This means once the technology with the highest variable costs, referred to as the marginal technology, has set the price, all other technologies will earn so-called infra-marginal rents (the difference between their own variable costs and the price). These rents are not only legitimate but essential for the functioning of the market, since they serve for the financing of the high fixed costs of investment of power generation.[4] As spelt out in Chapters 6 and 7, the level of these infra-marginal rents ultimately determines the competitiveness of nuclear energy against coal- and gas-fired power generation.

Thus even in isolation, electricity markets pose a number of conceptual challenges. Introducing carbon pricing adds an additional layer of complexity to the cost and profitability calculations of private investors. First and foremost, of course, carbon pricing will enhance the profitability of low-carbon sources such as nuclear and renewable energy. In a second step, one needs to consider the impacts on price volatility, and hence risk, of a carbon trading system or a carbon tax. Depending on the interaction of electricity and carbon prices, factoring the latter into an operator's profitability calculation may either smooth his/her stream of profits per MWh or render it more volatile. The results of this study in Chapter 5 indicate the latter, which means that other things being equal, nuclear operators have nothing to fear from carbon trading.

Third, the interaction of electricity and carbon pricing in a liberalised electricity market can have significant unforeseen side-effects. The most striking example of this is that under certain assumptions very high carbon prices of EUR 70 or more can increase the competitiveness of gas against nuclear. This is due to the fact that carbon-intensive coal then sets the electricity price boosting the profits of both gas and nuclear, but gas increases its profits per unit of output in this scenario at a faster rate than nuclear. Of course, the calculations provide again more intuitive results once one assumes that coal-fired power plants will be equipped for CCS, in which case high carbon prices will unequivocally benefit nuclear energy.

---

3. This effect is analysed and quantified in terms of a "suspension option" during the course of the study. See Chapters 4, 5 and 6.

4. For a complete exposition of the formation of infra-marginal rents see Joskow (2006) and Keppler and Cruciani (2010).

The previous point underlines that in real-world electricity markets the competitiveness of one technology depends heavily on developments in other technologies, an interaction that is wholly absent from LCOE calculations. Another example of this interdependence is the level of gas prices, which paradoxically affects the profitability of nuclear far more than that of gas itself. The profitability of gas, which is frequently the marginal fuel and thus determines electricity prices is in itself relatively immune to changes in the level of gas prices. However, every drop in gas prices bites into the infra-marginal rents that nuclear relies on in order to finance its fixed costs.

Subsequent chapters will develop these aspects in detail. There is, however, one general point worth highlighting already. By and large, one can consider investment decisions based on LCOE calculations as socially optimal, while investment decisions taking into account price risk as well as the specific price formation mechanisms in electricity markets will be privately optimal for an individual investor. Liberalised electricity market can thus create a wedge between socially and privately optimal objectives that may be of relevance to policy makers.[5]

This is particularly relevant for policy makers interested in the competitiveness of nuclear energy. As a technology with high fixed costs that need to be recuperated over long lifetimes, nuclear energy is particularly vulnerable to electricity price risk. This vulnerability is over and above that of its coal- or gas-based competitors, which benefit from the ability to suspend and defer production when prices are low.[6] It is thus not too much to say that nuclear energy is disproportionally affected by the switch from regulated to liberalised electricity markets. The study shows that this effect is noticeable but not dominant. In other words, even though nuclear energy is affected by volatile prices in liberalised markets more than other technologies, the usual determinants of profitability such as overnight costs and the cost of capital will ultimately play a larger role. Nevertheless, the link between relative competitiveness and institutional set-up is an issue for policy makers to remain aware of.

1.3 Scope of this study

The present study will examine these different issues in the following order. Chapter 2 provides some background on the institutional set-up for carbon pricing and the data used and also includes a first discussion of key issues such as the value of flexibility in investment. Chapter 3 consists of a review of the research on the issue of carbon pricing and competition in electricity markets. Chapter 4 presents a number of sensitivity analyses with respect to carbon pricing on the basis of the LCOE calculations of *Projected Costs of Generating Electricity*. Chapter 5 contains the "profit analysis", the assessment of the profitability of different existing power generation technologies in European electricity markets during the past five years. Chapter 6 includes an extensive analysis of the profitability of new green-field investments in power generation under the assumption that the future price environment will resemble the recent past. Chapter 7 assesses the evolution of the profitability of nuclear, coal and gas under different carbon price scenarios and Chapter 8 draws policy implications of the different results and concludes.

---

5.   One should not infer from this that liberalised electricity markets are necessarily inferior to regulated markets in terms of social welfare. The bounded rationality and "capture" of regulators on the one hand, as well as the dynamic benefits of liberalised markets such as new services, technologies and organisational forms on the other, may well outweigh any divergence in terms of the static welfare of market approaches resulting from considerations of social or private optimisation.

6.   This is not a technical but an economic argument. Even if nuclear power plants were technically able to switch production on and off at will and at no cost, they would not do it most of the times because their marginal costs would still be lower than prices. Nevertheless, their profitability would be penalised every time prices fell below average cost, since they could not adequately repay their fixed costs. In other words, at low prices a nuclear power plant would still gain money on each MWh but not enough to repay investment costs fully.

Chapter 2

Background

This chapter provides background, context and perspective for the later chapters presenting the actual results of this study. It shows the contribution of carbon emissions from fossil-fuel combustion in OECD countries and, in particular, the power sector, to global greenhouse gas emissions and gives some information on current carbon pricing initiatives in OECD countries. While the different methodologies employed will be presented in detail in Chapters 5 to 7, a brief discussion of different methodologies for assessing the competitiveness of nuclear energy will also introduce them below to see their complementary nature in context. Similarly the sections on the "suspension option" and the "scenario analysis" will explain the function of key building blocks, while leaving the technical details for later chapters. Finally, a sub-section will comment on the different data sources used.

2.1 CO_2 emissions from power generation and carbon trading

The pricing of climate change-inducing greenhouse gas emissions has increasingly become a reality in industrialised countries trying to reach their emission reduction targets defined under the 1997 Kyoto Protocol. Given that CO_2 emissions, frequently referred to as "carbon emissions", constitute the largest and most easily measurable share of greenhouse gas emissions (76% of the global total), it is no surprise that emission reduction efforts have been primarily concentrated in this area.[1] Of these 79% are due to the burning of fossil fuels and of these roughly 40% are due to the generation of electricity and heat in the power sector, where the burning of coal contributes about three quarters of all carbon emissions.

The power sector has two additional features that make it an attractive first target for emission reduction efforts. First, due to their high costs of transport, electricity and heat are produced for domestic or regional markets and are thus largely isolated from international competitive pressures. Second, the demand for electricity is highly inelastic, in particular as far as residential uses are concerned. This means that any additional costs due to CO_2 emission reduction efforts can be easily passed on to customers. While this may raise distributional issues, it has from an overall economic perspective the advantage of not radically affecting existing production and consumption patterns.

---

1.   According to *CO_2 Emissions from Fuel Combustion* (IEA, 2010a), 76% of global greenhouse gas emissions were due in 2005 to CO_2 (measured on the basis of their global warming potential in terms of CO_2 equivalence). Of these 79% (60% of the total) were due to the burning of fossil fuels, the remaining CO_2 emissions being mainly due to deforestation and land-use change. The next important greenhouse gas is methane (CH_4) which has contributed 16% of total, global greenhouse gas emissions.

   In 2008, the world emitted roughly 49 billion CO_2 equivalent tonnes of greenhouse gases. Of these, 29.4 billion tonnes (precisely 60%) were emitted world wide due to fossil fuel combustion. OECD countries contributed 12.6 billion tonnes or 43% of the global total of CO_2 emissions due to fuel combustion. From a sectoral point of view, CO_2 emissions from "electricity and heat production from both main activity producers and auto-producers", in short the power sector, contributed 12 billion tonnes (41% of total emissions from fuel combustion). Five billion tonnes were emitted by the power sector in OECD countries, of which 3.7 billion tonnes can be attributed to coal, 0.3 billion tonnes to oil and 1 billion tonnes to gas.

The precise macroeconomic impact of carbon taxes would require a much more extensive treatment. However, contrary to traditional taxation for revenue generation, environmental taxation such as carbon pricing geared towards the internalisation of externalities does not create the usual "deadweight" efficiency loss in overall welfare terms. However, this does not preclude environmental taxation causing efficiency losses in economic and financial terms, which would be reflected in GDP indicators. The issue is further complicated by the fact that environmental taxation sets in motion a number of dynamic effects, with both negative and positive impacts on economic growth, ranging from the delocalisation of polluting industries to the creation of new "green" industries. The OECD "Green Growth" project discusses many of these issues in far greater detail than would be possible in this context.

Carbon pricing is thus usually constructed around the power sector. Carbon pricing is already a reality in several OECD countries (see Box 2.1).

Box 2.1: Emissions trading in OECD countries	
European Union (EU) ETS 2005-2012 Free allocation 2013- Auctioning	Energy and industrial sectors, aviation from 2012. Approximately 11 500 installations covered. Installations >20 MWh combustion, specific production thresholds for industrial processes.
Switzerland 2008-	Voluntary participation by energy intensive industries that negotiate exemption from CO_2 levy. Approximately 350 companies.
New South Wales (Australia) 2003-	Electricity sector only. Electricity generators, retailers. Large consumers (>100 GWh p.a.) may choose to manage their own obligations.
US Regional Greenhouse Gas Initiative (RGGI, Northwest US) 2009-	Electricity sector only. Generators >25 MW capacity.
Alberta 2007-	Electricity and industry (oil sand mines, coal power plants). Large emitters >100 000 tonnes per annum.
New Zealand 2008-	Economy-wide once fully phased in: energy, transport, industry, waste, forestry, agriculture. Industry-specific thresholds for participation.
Tokyo Metropolitan 2010-	Commercial buildings and factories. Sites >1 500 kl of oil equivalent per annum (ca 1 400 sites).
UK CRC Energy Efficiency Scheme 2010-	Large businesses and organisations not covered by EU ETS. Organisations using >6 000 MWh electricity.
Western Climate Initiative (CA, NM, British Columbia, Ontario, Quebec) 2012-	Covers energy, industrial, liquid fuels sectors, depending on decisions of individual states. Emissions threshold >25 000 tonnes per annum.
Australia CPRS On hold	Energy, transport, industry, waste; opt-in for afforestation. Ca 1 000 sites >25 000 tonnes per annum.
USA H.R. 2454 (Waxman Markey) On hold	Energy, industrial, liquid fuels sectors; agriculture, forestry, waste not included. Ca 7 400 sites >25 000 tonnes per annum.

Source: Adapted from IEA, 2010b.

In addition there exist several carbon trading schemes under development that are likely to become a reality in the near future either in other OECD countries (Japan and the Republic of Korea) or key non-OECD countries (Brazil, China). Of course, a number of countries have also introduced carbon taxes. Levels vary but are usually far less than the average price, currently around 15 EUR/tCO$_2$, seen in the European Emissions Trading System (EU ETS), the most important of the existing carbon trading schemes. Among the OECD countries that have introduced carbon taxes are Denmark, Finland, Ireland, the Netherlands, Norway, the Republic of Korea, Sweden and the United Kingdom. A number of provinces and states in both Canada and the United States have also introduced carbon taxes.

2.2 Key functions and forms of carbon pricing

Naturally, carbon pricing has an impact on the competitiveness of different forms of power generation, enhancing the competitiveness of power generation from low-carbon sources such as nuclear energy and renewables to the detriment of fossil-fuel-based high-carbon sources such as coal and gas. The purpose of this study is to assess the precise form of this impact under real market conditions for different forms of carbon pricing and realistic assumptions about key parameters such as discount rates and fuel prices. This study thus refines the results of the IEA/NEA study *Projected Costs of Generating Electricity: 2010 Edition* by concentrating on the competitiveness impacts of carbon pricing for dispatchable baseload generation, that is nuclear energy, coal and gas.

Comparisons on the basis of the LCOE calculations of the kind performed in the Projected Costs study are primarily useful for an industry environment without price and bankruptcy risk, i.e., an environment of rate-of-return regulated utilities. This new study instead reflects the fact that an increasing number of utilities in OECD countries are now working in liberalised electricity markets with volatile prices and a non-negligible risk of bankruptcy. This means that two additional criteria need to be taken into account. First, with the possibility of prices falling below marginal cost, the option for producers to suspend or rather to defer production needs to be taken into account. This issue is discussed below under the heading of "suspension option". Second, the probability for an investment to make lower than expected profits or losses needs to be taken into account in addition to average expected profits. This issue is discussed below under the heading of "scenario analysis".

In addition, not all forms of carbon pricing have identical impacts on the relative competitiveness of nuclear energy and fossil fuels even at nominally identical average carbon prices. One can distinguish three major forms of carbon pricing:
1. carbon emissions trading with free allocation of permits ("grandfathering");
2. carbon emissions trading with auctioning of permits; and
3. a flat tax on carbon emissions.

Option 1 is distinguished from options 2 and 3 by the fact that it leaves substantial surplus profits (rents) with operators of fossil-fuel plants. This means that while carbon prices affect the merit order between nuclear and fossil-fuel plants, not only nuclear energy but also fossil-fuel-based production will now generate far *higher* profits than in the absence of carbon pricing. This is due to the integration of the opportunity costs of carbon permits into the price of electricity. In assessing the difference for operators between a permit system with and without free allocation, the study is particularly timely in view of the switch of the EU ETS, the most important of the carbon trading schemes, to full auctioning in the electricity sector in Phase III beginning in 2013.

Option 3 is distinguished from options 1 and 2 by the *stability of the cost differential* it imposes on operators. Depending on its correlation with other variables, the stability of the carbon price will affect the stability of the profit stream for fossil fuels and thus their relative competitiveness compared to nuclear energy. The study will show that the price volatility of carbon trading can indeed *positively* affect the competitiveness of nuclear energy.

Option 2 is, of course, the most interesting one as carbon pricing introduces an additional source of risk for carbon emitting technologies based on fossil fuels. To the extent that carbon prices are uncorrelated with coal and gas prices, volatile carbon prices in a trading system will increase the volatility of the revenues of the operators of coal- and gas-fired power plants. In such a case, nuclear competitiveness will benefit more from carbon trading than from a carbon tax.

A directly related issue is the volatility of electricity prices, which in a deregulated electricity market maintains a complex relationship with both carbon and fuel prices. With regulated electricity prices, the only volatility stems from carbon and fuel prices. Because nuclear is relatively insensitive to fuel cost, and it is unaffected by carbon pricing, its revenue and profits would be stable under this scenario, whereas the revenue of fossil fuels plants would not be. Stable electricity prices, either prices in the form of regulated prices or in the form of long-term contracts, will always enhance the competitiveness of nuclear energy relative to coal- and gas-fired power generation.

In deregulated electricity markets with liberalised electricity prices, the key question is the extent to which carbon prices are correlated with electricity prices and if the former increase or decrease the latter's volatility. If carbon prices and electricity prices (and thus gas prices) are highly correlated, the profits of the operators of gas- and coal-fired power plants may become less volatile and a carbon trading system could negatively affect the competitiveness of nuclear compared with a carbon tax. The empirical evidence based on historic correlations, however, points into the opposite direction: carbon trading does not smooth the profit stream of fossil-fuel-based generators and the competitiveness of nuclear energy increases more strongly with a carbon trading system than with a carbon tax.

2.3 Three different methodologies for assessing the competitiveness of nuclear energy

Throughout the study, the competitiveness of nuclear energy is analysed in three distinct ways, which are presented briefly in the following. Complete results for each methodology are subsequently provided in Chapters 4 to 7.

LCOE sensitivity analysis on the basis of Projected Costs study

The first case is based on the methodology and the data of the Projected Costs study. On the basis of the LCOE results obtained, an extensive range of sensitivity analyses are performed with respect to both discount rates and carbon prices. Carbon prices in this case are assumed to come in the form of a flat tax that reflects the political consensus on the social costs of carbon emissions. As mentioned above, this yields the comparative social resource cost of different power generation technologies and is thus an important indicator for policy makers and modellers. It is also a valid measure of the investment and operating costs of different power generation technologies from the point of view of a utility operating in an environment where both carbon electricity prices are regulated and solvency risk is absent. These results on the basis of the LCOE methodology are presented in Chapter 4.

To the extent, however, that the utilities operate in liberalised electricity markets, with carbon prices set either in an emissions trading market or as a tax, two different methodologies need to be employed to adequately reflect the risk-reward structure of the, frequently private, investors deciding among the different possibilities for power generation. These two methodologies are distinguished throughout the study as "profit analysis" and "investment analysis". Both are based on the reported market data generated over the past five years in the European energy markets for electricity, coal and gas as well as the EU ETS for CO_2 permits.

Profit analysis

The profit analysis estimates the impact of introducing a carbon price on the profits of the coal, gas and nuclear power generations for *already existing* power generation plants. It is thus a backward-looking *ex post* analysis of the short-term effects of different carbon regimes based on historic data for electricity, carbon, coal and gas prices. Profit analysis thus only considers the *variable costs* associated with running existing power plants, while the investment costs of building the plant are not accounted for. While this may seem to reflect only a very partial segment of the economic reality of operators, the profit analysis nevertheless reflects their true experience since the introduction of the EU ETS. The base case of the profit analysis presented in Chapter 5 thus provides an indication of the real profits made by the operators of different power plants during the 2005-10 period.

This base case, which keeps as closely as possible to the observed reality of carbon pricing with a costless allocation of permits, is contrasted with two alternative cases, which both imply actual payment for carbon emissions. In the first of these alternative cases, carbon prices observed in the EU ETS are maintained but their acquisition is imputed as a cost (as would be the case, for instance, in a continuous auction). In the second of these two cases, carbon prices are substituted by a flat carbon tax corresponding to the average price during the 2005-10 period. The profit analysis thus allows assessment of two different issues:

1. showing how true carbon pricing would impact the level and volatility of profits from electricity production; and

2. providing a rough but robust method for comparing the impact of the two different carbon pricing regimes (EU ETS or equivalent carbon tax) on the competitiveness of different power generation technologies from a short-term perspective.

The different carbon pricing regimes are compared both with respect to the level of the per unit profits obtained by nuclear, coal- and gas-based electricity generators as well as with respect to the volatility of the per unit profits. The two parameters are integrated with the help of the Sharpe ratio (see Chapter 5 for a detailed discussion), which provides a risk-adjusted measure of profitability. Comparing thus both the level as well as the volatility of profits in one single measure allows, in particular, to evaluate the impact on competitiveness of a carbon trading system with that of a carbon tax.

Investment analysis and carbon tax analysis

The investment analysis is perhaps the most substantive contribution of this study. It aims at determining the relative competitiveness of different technologies for new, yet-to-be-built power plants under different assumptions for carbon prices. It thus takes a forward-looking long-term view. A first scenario models the relative profitability of power plants based on nuclear, coal and gas under the assumption of a carbon price of EUR 14 per tonne of CO_2 (the average price in the EU ETS during the 2005-10 period). Other than assumptions about carbon prices, assumptions concerning the level and structure of electricity prices are of key importance. The NEA carbon pricing study employs the assumption that the level and the structure of electricity prices will be identical to those prevailing throughout the 2005-10 period. For a plant with a lifetime of 40 years, the study thus reproduces eight times the price dynamics (including the assumptions and correlations for fuel prices) of the past five years. Clearly, this is a rather audacious assumption. However, in contrast to any explicit modelisation of electricity prices its great merits are its transparency and the absence of any modelling bias.

In addition to the observed prices over the past five years ("base case scenario"), the study also contains a "high price case" and a "low price case" scenario. The first reproduces the price and cost dynamics of the 12 months during the 2005-10 period where electricity prices were highest and the second considers the 12 months during which they were lowest. The differences in relative profitability between the different technologies in the three cases are instructive and show the importance of electricity price assumptions in addition to those for carbon prices.

In addition to the scenario of EUR 14 per tonne of CO_2, the study explores the relative profitability of nuclear, coal and carbon under a range of carbon prices reaching from EUR 0 to EUR 100 per tonne of CO_2 in the carbon tax analysis in Chapter 7.[2] Again, the results considerably add to simple intuition. While the level of carbon prices has a strong negative impact on the relative profitability of coal, which is to be expected, it only has a paradoxical impact on the relative profitability between nuclear and gas. This is due to the fact that with high and very high carbon prices, electricity prices are set by coal, which allows gas to earn additional infra-marginal rents, which largely off-sets its own increased carbon costs. Indeed the analysis shows that under such circumstances gas prices are likely to be a more important determinant of the competitiveness between nuclear and gas than carbon prices. There exists thus a "window of opportunity" for carbon prices between EUR 20 and EUR 50 per tonne of CO_2 where their impact on the competitiveness of nuclear power is greatest.

A key question for the investment analysis was choosing the appropriate measure of the relative profitability of the different technologies, the best known measures of profitability being the net present value (NPV) and the internal rate of return (IRR). Both measures, however, have drawbacks. Calculations of NPV, which is the sum of the discounted flow of all income and expenditure, clearly favour large projects over smaller ones. On this measure, even an only marginally profitable nuclear plant could, due to its size, trump a smaller gas plant even if the latter was very profitable on a per unit basis. Pure NPV calculation would be an appropriate measure if only one single plant could be built at a given location and no alternative investment opportunities existed, which is clearly not the case in a large integrated electricity market, where investors will choose the investment with the highest return.

In principle, the measure of IRR avoids this pitfall by calculating the return on capital over the lifetime of the project. It does, however, have two major drawbacks of its own which make it unsuitable for the comparison of different technologies. First, IRR calculations assume that interim cash flows are immediately reinvested at the same rate as the one generated by the whole project, which are also equal to the assumed cost of capital. This is a somewhat unrealistic assumption especially for projects with high rates of return. The second even more important reason is that IRR calculations are not a very good means to assess the relative profitability of projects or technologies with different lifetimes. They are primarily a means to decide whether a given project should go ahead or not given an exogenously set hurdle rate defined by the opportunity cost of capital.

2.　For the impact of carbon prices on electricity prices, the study uses an assumption of 100% pass-through. This means that electricity prices vary in function of carbon prices from EUR 30 to EUR 130 per MWh.

Modified internal rates of return (MIRR) calculations do get around the first issue as they allow using independent parameters for the cost of capital and the reinvestment rate, in short the cost of borrowing and the benefits of lending. MIRR calculations also avoid the vexing problem of IRR calculations of producing multiple solutions in case of negative cash-flow after the initial investment (such as would be the case for waste disposal or decommissioning). Despite these advantages over IRR calculations, MIRR calculations remain a methodology for assessing a given project against an exogenously set opportunity cost rather than for comparing the profitability of different projects with different fixed costs and lifetimes.

With NPV not accounting for size and IRR and MIRR unsuited for comparisons between different technologies, the present study uses a modified measure of NPV that normalises for project size and provides a measure for the value that is created for investors over the lifetime of a project. This measure is called the profitability index (PI) and corresponds to the NPV normalised by investment costs:

PI = NPV/INV.

Both the net present value and investment costs are, of course properly discounted to the date of commissioning. Any viable project will thus generate a positive PI, which means that investors are not losing any money. In principle, one might have normalised NPV also over other parameters such as output over the lifetime of plant. Normalising by investment costs, however, means that the PI provides an answer to the question at the basis of this project: among nuclear, coal or gas, which one would generate the highest return on the investment of a private investor in a liberalised electricity market? The answer is, the one with the highest PI, once it is calculated as NPV normalised by investment costs.

2.4   Data and the EU Emissions Trading System

To the extent that the competitiveness of nuclear power is assessed on the basis of the interaction of carbon prices and electricity prices in a liberalised power markets, the present study uses data from the EU ETS. The EU ETS is the world's largest and best developed emissions trading system. It is also the only system for which there exist daily data on carbon prices for more than five years (see Box 2.2). In addition, Europe possesses a rapidly integrating electricity market. This allows the study of the interaction between carbon prices and electricity prices, all important for determining the competitiveness of nuclear power in liberalised electricity markets, in a real-world context.

Due to the availability of daily data for different variables, the study covers the period from July 2005 to May 2010 with a complete set of *daily* price data from European energy markets. Carbon prices are thus the spot prices for EU Allowances (EUAs) traded on the EU ETS provided by Bluenext, the largest European exchange for the spot trade in EUAs. The data for electricity, gas and coal prices also pertain to European energy markets. For gas prices, daily data from the Zeebrugge gas hub (Platt's day-ahead) were used. The data for coal and electricity prices were provided byEEX, the largest European electricity exchange operator. Daily coal prices pertain to month-ahead futures on the Rotterdam coal market (ARA coal). The story is slightly more complicated for electricity prices, where the daily prices used in the analyses of this study are an average of the prices for day-ahead spot delivery, monthly, quarterly and yearly forward contract, weighted by the respective daily volumes sold in each market segment.

The reason why this particular method was chosen is, of course, that electricity is a non-storable good and thus spot and forward prices are only very imperfectly correlated and can diverge widely. On the other hand, each MWh produced by a power plant at a given day contributes to its profitability as a function of the price that it is able to obtain whether on the spot or on the forward markets. Since the objective of this study is to provide a true measure of the profitability of different power technologies, the average over different time horizons is the appropriate measure. In the case of coal and gas, storage is also imperfect and costly. Nevertheless, some storage exists and spot and future prices are closely correlated. This justifies concentrating only on the most liquid market segment for coal (month-ahead) and gas (day-ahead). For carbon prices, finally, the question of which length of contract to use, does not arise. As a perfectly storable financial asset, the spot and forward prices for CO_2 diverge only by the rate of interest with near perfect correlation.

Box 2.2: The European Emissions Trading System (EU ETS)[3]

The EU ETS scheme covers medium and large emitters, including electricity generators, pulp and paper, steel and cement, producers with combustion facilities greater than 20 MW. As of 2010, around 11 000 facilities in 27 member states (as well as in Iceland, Liechtenstein and Norway) are included, covering 45% of European CO_2 emissions. Aviation is to be included from 2012, and aluminium production from 2013. Initially only carbon dioxide was covered, but from 2013 this is to be expanded to a number of other greenhouse gases produced.

The EU ETS overall cap is 6.5% below 2005 levels for the 2008-12 period and will decline to 21% below 2005 levels in 2020. The EU ETS began with a trial phase (Phase I) from 2005 to 2007, and is now in its first phase of full trading from 2008 to 2012 (Phase II). The most significant change concerning the 2013-20 Phase III concerns the auctioning of the emission permits that have hitherto been largely given out to emitters for free based on their historic emissions ("grandfathered"). Overall, more than 50% of permits will be auctioned from 2013, a share that will be increasing each year. In the electricity sector, however, 100% of emissions will be auctioned from the very start of Phase III.

In its short history, the EU ETS has already experienced dramatic price swings. Prices had climbed as high as EUR 30/tCO_2 in 2006 when the publication of the first year's audited emissions inventories revealed a surplus of allowances. In conjunction with the inability to "bank" allowances for future use, this over-allocation resulted in a decline towards a price of virtually zero at the end of Phase I. During Phase II, prices have evolved in a band of EUR 12 to EUR 18, with an average price of EUR 14 between 2005 and today. Phase II from 2008-12 was designed to coincide with the first commitment period of the Kyoto Protocol, and is a major mechanism for meeting Europe's Kyoto commitments. For the time being, surplus allowances due to the sharp drop in industrial output and power generation in 2008 and 2009 have not led to a price collapse since allowances can now be banked for use in Phase III. During 2009, 6 326 million tonnes of allowances were traded in the EU ETS, at a market value of USD 118 billion.

All technical data for the different technologies stem from the IEA/NEA study *Projected Costs of Generating Electricity: 2010 Edition*. They include the data for the costs of overnight investment, operation and management, the costs for waste disposal and decommissioning, as well as the nuclear fuel costs. In addition, the Projected Costs study provides data on load factors, carbon intensity and the efficiency of converting fossil fuels into electricity. In each case, the mean values for the entries provided by European OECD countries were used for this study.

---

3.   The information was drawn from IEA (2010b), Ellerman, Convery and de Perthuis (2010), European Commission (2010) and World Bank (2010).

A final, crucial, parameter to be defined is the cost of capital. As is well known, and borne out by the results of the Projected Costs study, the more capital-intensive a technology is the more sensitive it is to the cost of capital. Other things being equal, low-carbon technologies such as nuclear and renewables are more capital-intensive than fossil-fuel-based technologies, since they replace fuel costs with more sophisticated and expensive fixed investment. The Projected Costs study used 5% and 10% as the real cost of capital. Under the assumptions of the study, which included a carbon price of USD 30 per tonne of CO_2, nuclear was easily the most competitive source at a 5% discount rate, but with a 10% discount other technologies became more competitive for data from OECD Europe. (Nuclear remained the most competitive technology also at 10% for OECD Asia and OECD North America.)

The present study instead uses only one single real discount rate of 7%. In comparison nominal rates for long-term corporate bonds in the European utility sector are around 5%.[4] Given that the long-term inflation target of the European Central bank is "below but close to 2%", one can consider that the cost of debt for European utilities is currently at around 3% real. Of course, no utility would be able to rely entirely on debt financing to build a new nuclear power plant but would also need to rely on equity investors who may demand much higher rates, say, between 10% and 15% nominal, which corresponds to real rates between 8% and 13%. The precise ratio of debt and equity finance and the precise demands of equity investors would, of course, depend on the financing model that be used. The latter will include guarantees on regulatory procedures, licensing, the carbon policy and a host of other issues. Suffice it to say that a 7% real cost of capital in the current monetary environment is a rather conservative assumption from the point of view of nuclear power production.

2.5   The merits of flexibility and low fixed-cost-to-variable-cost ratios

In addition to their sensitivity to the rate of interest, technologies with relatively higher fixed-cost-to-variable-cost ratios, such as nuclear and renewables, have an additional disadvantage when switching from a regulated environment with guaranteed electricity prices to deregulated markets with volatile electricity prices. The change in the institutional framework has direct methodological implications for determining the relative competitiveness of different power generation options. In fact, an assessment of LCOE as was performed in the Projected Costs study is an appropriate methodology for regulated electricity markets. In fact, the result of the LCOE calculations yields the power price the regulator needs to ensure so that a given technology obtains a pre-set level of remuneration for its investment (the cost of capital that is assumed in the LCOE calculation).

In an environment of deregulated wholesale markets, such as the one prevailing in the European Union since 1997, utilities are exposed to volatile prices. It has been a regularly voiced criticism of the LCOE methodology that the regulated market environment for which it is primarily designed is found only in an ever smaller subset of OECD member countries. The publication of the influential 1994 book by Dixit and Pindyck on *Investing under Uncertainty*, which highlights the value of flexibility in investment in terms of "real valued options" (which are not captured by the LCOE methodology), has further incited analysts to pay attention to fixed-cost-to-variable-cost ratios. The answer to such criticisms of the LCOE methodology is precisely the use of alternative methodologies to assess profitability and relative competitiveness such as the profit analysis and the investment analysis

---

4.   It is instructive to look at publicly accessible sites for current rates of European corporate bonds in the utility sector such as www.comdirect.de/inf/anleihen/index.html. As of 28 January 2011, EDF corporate bonds had nominal rates between 4.6% and 5.4% for durations between 14 and 30 years. Bonds for Enel yielded around 5% for 12 years duration. RWE bonds yielded 5.2% for 22 years duration and Vattenfall bonds 4.5% for 13 years duration.

pursued in this study. These analyses confirm the intuition that technologies with relatively lower fixed-cost-to-variable-cost ratios have some advantages in liberalised electricity markets. However, the findings yield a surprisingly nuanced picture that investors need to recognise before drawing overall conclusions in too hasty a manner. In order to understand the different quantitative results in Chapters 5 and 6, one need to distinguish first the different mechanisms on a conceptual level.

A comparatively high fixed-cost-to-variable-cost ratio is a distinctive feature of low-carbon technologies for power generation. It thus holds for nuclear as it holds for renewables or coal-fired power generation with carbon capture and storage, and even for demand-side investments such as energy efficiency improvements. All are characterised by large up-front investments which must be recouped MWh by MWh over relatively long time frames. On the other hand, the relatively low fixed costs of fossil-fuel-based technologies are, of course, compensated by the high costs of the fossil fuels themselves, whether they be gas, coal or oil, as well as by the cost of the greenhouse gas emissions they generate. Putting a price on the carbon emissions from these fossil fuels is thus also a means to overcome the disadvantage of high fixed cost technologies in liberalised electricity markets. In addition to determining the relative impact of a carbon tax or a trading system on the competitiveness of nuclear energy, this study thus also aims at determining the height of the carbon value required to overcome the disadvantage of carbon-free, high fixed cost technologies in deregulated electricity markets.

Understanding the different impacts of the fixed-cost-to-variable-cost ratio

There are three different mechanisms by which the fixed-cost-to-variable-cost ratio impacts profits, each of which depends on different factors in the market. The ensuing explanations are perhaps easiest to understand if one applies them to two different technologies, say nuclear plants and combined cycle gas plants, which have very similar average costs over their respective lifetimes but different fixed-cost-to-variable-cost ratios. Roughly speaking, nuclear energy has a fixed-cost-to-variable-cost ratio of 2:1, investment costs are thus two-thirds of total costs, whereas natural gas has a fixed-cost-to-variable-cost ratio of 1:2 and fuel costs are thus two-thirds of total costs. The three mechanisms by which this difference makes itself felt are:

1. The greater ability of technologies with lower fixed-cost-to-variable-cost ratios (and consequently higher variable costs) to ride out transitory periods of low prices and thus make use of a "suspension option" (see Chapter 5 for results).

2. The relatively smaller lock-in for investors in the case of low fixed-cost-to-variable-cost ratios in the case of permanently lower than anticipated prices, i.e., at identical capacity a lower financial risk in the case of "stranded assets" (see Chapter 6 for results).

3. The ability of technologies with lower fixed-cost-to-variable-cost ratios (and consequently higher variable costs) to set electricity prices as the marginal fuel and thus to reduce the volatility of profit margins (see Chapters 5, 6 and 7 throughout).

In the following, we will discuss the three sources of advantages for low fixed cost (and consequently high variable cost) technologies and their relative merits. Before doing so, one needs, however, to understand also two major *disadvantages* of comparatively high variable cost technologies such as natural gas and coal in liberalised electricity and carbon markets. These two disadvantages are the following:

1. The volatility of the prices of the underlying fuel, natural gas or coal, is a source of volatility of profits even if either natural gas or coal is the price setting fuel. In fact, the volatility of profits during the 2005-10 period was *higher* for natural gas than for nuclear.[5] Despite volatile market prices for electricity, the cost stability of nuclear energy proved to be an advantage (see Chapter 5 for results).

2. The dependence of natural gas and coal on fossil fuels is at the root of an additional source of uncertainty in the context of carbon markets such as the EU ETS. Due to the high volatility of carbon prices also the volatility of the profit stream for gas and coal is increased (see again Chapter 5 for results). This volatility of carbon prices might even extent to a form of political risk in case a country suddenly decides to strengthen its carbon emissions policy.

Let us come back to the often mentioned but frequently ill understood advantages for a given technology of having low fixed costs in a liberalised power market with volatile prices that are set by the technology with the highest marginal cost.[6] Another way of expressing this is to consider that the high variable cost technology has a "real valued option" to suspend production, which a high fixed cost technology with lower variable costs does not possess or possesses to a much lower extent. Concerning this greater ability of technologies with lower fixed-cost-to-variable-cost ratios (and consequently higher variable costs) to ride out periods of low prices, this study provides clear insights. It has clearly identified and measured the value of this "suspension option" but has also established that its quantitative impact is limited, less than 18% of total cost under the most favourable circumstances.

The fact of having relatively higher variable costs can be compared to have a form of insurance against transitory lower prices, or in other terms to possess an option to defer spending on fuels until prices pick up (see Box 2.3). When prices fall below their expected level, gas turbines will stop producing. They will make zero profits, but will save on expensive gas in the process, which can be used later; payments for the relatively low fixed costs are limited. Nuclear will instead keep on producing and will even continue to make small profits. However, the bulk of its cost – the original investment – has already been expended and the clock to repay it is ticking. Due to its lower variable costs, nuclear power does not possess a suspension option and lower than expected prices will fully feed through to its profit calculations.

---

5. Volatility is measured throughout the study in terms of the standard deviation during the five-year period from July 2005 to May 2010.

6. The following explanations can also be read as an argument in favour of the observation that nuclear energy is better served by a regulated market with stable and predictable prices. There is no doubt that nuclear energy (as well as other low-carbon, high fixed cost technologies) are penalised by volatile and uncertain prices. This means liberalised electricity markets can lead to disconnect between private and social optimality. In other words, it is easy to imagine that nuclear (or other low-carbon technologies such as renewables) is the technology with the lowest average lifetime costs but that it will not be adopted by private investors who fear price volatility. In the case of renewables, this issue has been circumvented with the help of feed-in tariffs, which are, of course, a form of regulated prices. There is indeed an intellectually coherent case to be made that regulating prices in electricity markets can improve social welfare in static terms and avoid an otherwise carbon-intensive and more expensive generation mix.

Box 2.3: The value of an option to suspend production

The NEA study models the "suspension option" simply as an option to suspend production when prices fall below variable cost. This is a quite frequent configuration, in particular for gas, where shutting down and restarting production can be done at little extra cost. In the case of nuclear and coal, "ramp costs" need to be considered, as stopping and re-starting increase operating costs and may reduce plant lifetime. Precise information on "ramp costs" and the ability for load following is scarce, which is why the NEA has initiated a study on "system effects" to explore the question in detail. However, "ramp costs" are certainly of an order of magnitude lower than other costs and may be omitted in first approximation, in particular as there have been few instances in which the variable costs of nuclear or coal exceeded the synthetic price (a weighted average of spot and forward prices) in this study. The issue is different for the spot market, where, for instance, Germany has experienced more than 20 hours of *negative prices* during the past 18 months as large amounts of intermittent wind-power overload the system. Since wind-power is not remunerated through the market but through subsidised feed-in tariffs, wind-based producers have no incentive to leave the market even at negative prices.

The suspension option thus has value as it avoids situations when prices do not cover marginal cost. Behind this very intuitive and moderately important concept lurks a far more important but less intuitive one: with the suspension option an operator will be much more likely to encounter prices covering *average cost*. The suspension option is thus far more valuable for a high variable cost technology such as gas. First, the probability that it may be exercised is much higher. Second, since the difference between variable costs and average costs is much smaller, the chance that future prices above variable costs will cover or exceed average costs is much greater than for technologies such as nuclear.

Technically speaking, a suspension option allows capturing the "value of waiting for future information" that has been made familiar in *Investing under Uncertainty* by Dixit and Pindyck (1994). In their terminology, an investor in a high variable cost technology possesses a "real valued option" to wait for future price information. The *option to suspend production* is thus an *option to defer spending* on costs until prices are right. Consider the hypothetical case of a technology with zero fixed costs and high variable costs: even in a volatile market it will incur zero price risk. It will operate when prices are high and suspend production when prices are low. On the contrary, say, a renewable technology where all costs are fixed costs will be helpless in the face of price volatility. The greater the price volatility, the greater will be the value of the suspension option and the advantage of high variable cost technologies over high fixed cost technologies. This underlines once more that the competitiveness of high fixed cost technologies benefits from stable and predictable price environments.

The second point is a logical extension of the first and considers the case that prices would drop to levels that are permanently lower than expected, say substantially below the MWh cost of natural gas. Again gas would stop producing and, in case that there is no hope of prices coming back, go out of business altogether having to write off its initial capital investment. In this case, nuclear energy would continue producing since prices are likely to remain above its marginal cost, although investors will no longer be able to recoup their fixed costs.

Paradoxically, this situation is worse for an investor in nuclear power than for gas, even though the former continues to produce at a profit and the latter has left the market. How can this be? The investor in gas is writing off his/her initial, relatively modest, investment and is leaving the market with a small loss. The investor in nuclear will continue producing without any hope of ever recouping the totality of his/her initial investment and the uncovered share of his/her investment might well be larger than the total fixed cost of the gas turbine. For an investor having to make the choice between nuclear and gas, the probability of a prolonged period of low prices will thus be uppermost on his/her mind. A scenario analysis in Chapter 5 proves this point empirically for different probabilities of high and low electricity price scenarios. The higher the probability of a low price scenario, the more an investor will lean towards gas in order to protect the downside of his/her investment. However, if the probability of a permanent decline in prices due to overcapacity is low, the outlook for nuclear energy is very good. Quite obviously, the outlook is best, if the outlook for future price is certain: that is in a market where prices are regulated.[7]

The third consideration is slightly different from the other two as it concerns the relative profitability of gas and nuclear even if the average price of electricity does not vary over time and stays permanently above the marginal cost of gas. In this case, the advantage of high variable cost technologies is that they set the price. In other words, even with unstable prices their profitability is supposedly more constant, as their costs per MWh and their revenue vary in parallel. In contrast, a technology with low and stable variable costs such as nuclear energy would be exposed to a more volatile profit flow as the difference between costs and revenue would vary through time.

In theory, this argument is widely acceptable and it is also consistent with standard microeconomic theory. In practice, however, this study shows (see Chapter 5) that the volatility of profit flows is *higher* for power generation from gas than from nuclear energy. Reasons for this divergence from simple theory are that gas prices can be even more volatile than prices for electricity production (a substantial part of which is locked in through forward contracts), the added volatility from carbon prices during the 2005-10 period and varying profit margins due to variations in demand, which is influenced, among other things by largely unpredictable weather patterns.

In summary, there is some merit to the often cited argument that comparatively low fixed costs can improve the competitiveness of technologies such as natural gas against high fixed cost technologies such as nuclear. Yet, the empirical evidence hints at a more complex picture:

1. The ability of high variable cost technologies to draw on a suspension option and to skirt short-term decreases in electricity prices is real but small for historic price series. It might increase somewhat if electricity prices became considerably more volatile.[8]

2. The ability to limit downside risk in the case of permanently lower electricity prices is also real but depends entirely on the probability of such a shift happening. The advantage vanishes as the risk of a market collapse decreases.[9]

---

7.   Decision makers have at least intuitively grasped the fact that price uncertainty discourages investment in high fixed cost technologies and are experimenting with financing models *hors marché*. In the case of renewable energies, feed-in tariffs are the norm. In the case of nuclear energy, two innovative models, both of which involve large-scale consumers more directly, merit particular attention. The construction of the Finnish Olkiluoto reactor is thus financed by consumers, who have arranged with the operator – of which they are also majority shareholders – to buy electricity at average cost. The French Exeltium consortium instead organises a 20-year power purchase agreement at a fixed price between EDF and a number of electro-intensive industries.

8.   There is some evidence that prices in European electricity markets will become more volatile due to intermittent renewables such as wind-power. However, increased volatility in spot prices does not necessarily translate into increased volatility in the forward market. Especially prices for the dominant one-year forward contract, the calendar at which two-thirds of all registered market transactions are made, might be relatively unaffected.

9.   While the future is unwritten, one would have difficulties finding an expert who would consider the probability of such a long-term market collapse as very high during the next 20 years in OECD countries and *a fortiori* in non-OECD countries. Rising electricity demand due to increasing growth and a switch from less versatile energy sources (i.e., the direct burning of fossil fuels) on the hand and increasing difficulties to build new power generation projects due to the NIMBY syndrome on the other would presage a tightening of the demand and supply balance rather than the opposite. The inelasticity of both power supply and demand also makes operators in liberalised electricity markets fear overinvestment far more than underinvestment, which structures their investment decisions. Add to this structurally rising prices for CO_2 emissions as well as fossil fuels and the likelihood that electricity prices in the future will be lower than today is very unlikely indeed.

3. The ability of the marginal fuel to protect its profit stream by having the prices of output (electricity) vary with the price of its major input (natural gas) is not borne out by empirical analysis which finds the volatility of the profits of gas higher than the volatility of the profits of nuclear.

Overall, there remains a small benefit from having comparatively lower fixed costs and comparatively higher variable costs at equal average lifetime costs. However, the final outcome is dominated by questions of the absolute level of electricity prices, the margins of electricity prices above marginal cost as well as carbon prices rather than by fixed-cost-to-variable-cost ratios. In standard economic profit and loss analysis, abstracting for the moment from credit constraints and issues of political acceptance or security of supply, from the point of view of nuclear energy its high fixed-cost-to-variable-cost ratio is at the level of an inconvenience rather than a decisive competitive handicap even in liberalised markets. Average discounted lifetime costs remain key for establishing the competitiveness of different technologies. Although this study has strictly adopted the point of view of a private investor in a liberalised market, its results based on a measure of NPV normalised by investment size complement and corroborate rather than contradict the results of the LCOE analysis in the *Projected Costs of Generating Electricity: 2010 Edition*.

Chapter 3
Existing research on carbon pricing

Carbon pricing in power markets, either in the form of a tax or of an emissions trading system, naturally impacts the competitiveness of nuclear energy *vis-à-vis* fossil fuels. As a source of electricity without carbon emissions, the competitiveness of nuclear benefits from carbon pricing, in particular when compared with coal.[1] Both a carbon tax and an ETS affect the short-term profitability of different power generation options as well as the long-term investment decisions of operators. In addition, carbon pricing sends a strong dynamic signal for the realignment of R&D efforts and thus for future technology trajectories.[2]

Given the importance and complexity of the interactions between different forms of carbon pricing, the competitiveness of different technologies and investment decisions in the power sector, it is not surprising that a wide and varied theoretical literature has developed to study the phenomenon. Of course, this study has developed its own original approach in assessing the impact of carbon pricing on the competitiveness of nuclear energy. The major new contribution of this NEA study is providing an empirical evaluation of the impact of the EU ETS on the competitiveness of nuclear power. But this approach is informed by and built on the existing literature and this chapter provides an overview of this literature and presents its main strands.

3.1 Five distinct approaches in a wide and varied literature

When looking at the wide and varied literature on carbon pricing and nuclear energy, one can distinguish five major approaches, which can be briefly characterised as follows:

1. Some studies use a *profit analysis*. This approach is suitable to assess the performance of an existing power plant under different carbon price regimes, as it only considers variable costs and benefits without including any costs related to construction.

2. The studies using a *basic cash flow analysis* estimate either the NPV or the LCOE of a new investment. In this method, the investor compares the sum of *all* discounted costs and revenues of the investment.

---

1. This assumes that carbon emitters actually pay for their emissions. In emissions trading schemes where emissions permits are distributed for free on the basis of historic emissions ("grandfathering"), the impacts on competitiveness are less evident (see Burtraw and Palmer, 2007). On the one hand, operators may switch in the short term between generation options with different carbon intensities according to the carbon price, typically between coal and gas during periods of base-load when both are available. On the other hand, the impact on investment is less clear as carbon-intensive power generation options receiving their emissions permits for free benefit from the higher electricity prices generated by carbon pricing.

2. Such "price-induced technological change" (Hicks, 1932) is not the immediate focus of this project that remains confined to a methodology of comparative statics. Nevertheless, this effect should not be underestimated. The often evoked and yet to be realised technological milestones of commercially competitive renewable energy or industrial-scale carbon capture and storage (CCS) rely entirely in such price-induced technological change for their realisation.

3. A third group of studies assesses the impact of price volatility using *real option analysis*. A real option implies that the investor has the possibility (but not the obligation) to undertake certain business decisions like making, suspending or abandoning an investment. The real option is important in the presence of uncertainties when the investor does not know *a priori* the best investment decision to take.

4. Fourth, some studies do not consider investing only in a single power plant, but contemplate the possibility of having a portfolio of different plants (*portfolio analysis*). Here they calculate the share of nuclear power plants in an optimal portfolio.

5. Last but not least, there are a number of studies that prepare the necessary groundwork for subsequent analytical work by providing the necessary background information on the EU ETS (*ETS analysis*). This may include the synthesis of widely scattered statistical evidence, the careful analysis of institutional mechanisms or the identification of the causal relationships between the many cost and price variables that interact with carbon prices.

Below each method is briefly presented followed by a review of recent works.

3.2 Profit analysis

Under profit analysis we classify the studies that appraise how carbon pricing affects the profits of an existing power plant considering only current revenues and costs, i.e., variable costs or the costs necessary for running the power plant. Capital costs, the costs necessary to build the power plant, are considered as sunk and do not influence the analysis.

Green (2008) thus studies the profits of coal, gas and nuclear power plants under carbon taxes and carbon permits. The prices of energy and carbon permits are calculated using a supply and demand model that take into account volatility and correlations in fuel prices. According to this study, a carbon tax would increase the competitiveness of nuclear more than an ETS because a fixed price for carbon reduces revenue volatilities of nuclear generators and raises the revenue volatilities of fossil fuel stations.

Keppler and Cruciani (2010) use a profit analysis to assess the impact of carbon pricing on infra-marginal rents in the EU ETS Phase I due to the pass-through of the price of allowances received for free. They also assess the effect on rents due to switching from free allocation to auctioning. In their analysis, the rents generated during the first phase of the EU ETS are in excess of EUR 19 billion per year for electricity producers with carbon-intensive power producers gaining most due to the free allocation of allowances. With auctioning, carbon-free producers, such as nuclear power plants, will continue to benefit from increased infra-marginal rents due to higher electricity prices. Carbon-intensive producers instead will have to pay for allowances and will face substantial losses in comparison with free allocation.

In this study, the profit analysis is undertaken in two steps. In a first step, a profit analysis for coal, gas and nuclear power generations based on historic data from the EU ETS is performed to estimate the profits per unit of output for different generating technologies. In a second step, these profits are normalised for their volatility by calculating their respective Sharpe ratios. The Sharpe ratio is defined as the ratio of the profits of an asset and its standard deviation, the most basic measure of volatility, over time and provides a risk-adjusted measure of the profits. By comparing their respective Sharpe ratios one can then determine the respective impacts on profits of a carbon trading system and a carbon tax.

3.3 Basic cash flow analysis

There are several methods in finance to evaluate when a new investment is worth pursuing. The technique by far preferred by experts is the NPV analysis. The NPV is the sum of the present value of all the cash flows that occur in the project; thus it takes into account all the costs and revenues of the project during all its lifetime, discounted by the rate of return the investor is willing to apply in order to undertake the risks of the investment. The "NPV rule" says that any investment with a positive NPV is a good investment, and in the case of mutually exclusive investment opportunities, the investor has to prefer the project with the highest NPV (Brealey, Allen and Myers, 2006).

Another methodology that is often used to compare different power generation technologies and that is based on the cash flow approach is the LCOE calculation, which establishes the average price of electricity that would make the NPV of a new project equal to zero. The LCOE is equivalent to the price that would have to be paid by consumers to repay the investors for all the costs occurred in the project, discounted by the appropriate discount rate. The LCOE methodology has the limitation that it assumes a constant price of electricity and thus does not take issues of price risk or volatility into account.

In their basic versions, NPV and LCOE calculations do not explicitly account for price volatility since all risk is only considered in terms of the discount rate, which is usually determined by the cost of capital. They thus do not account for the specific uncertainties of different power generation investments (see Brealey, Allen and Myers, 2006).[3] In addition, in LCOE analysis the price of electricity is an output that is considered constant over time. This means the electricity price is not correlated with the prices of inputs as would be the case in a liberalised electricity market. These limitations make LCOE analysis more adapted for assessing power investment in a regulated market, where the price electricity is constant. To overcome these limitations, NPV (or LCOE) are often calculated under different scenarios from which readers can draw their own conclusions.

The Projected Costs study (IEA/NEA, 2010) provides the most recent LCOE calculations. It calculates the LCOE for almost 200 new or planned power plants in 21 countries. This study considers a fixed carbon tax of USD 30 per tonne of CO_2 with sensitivity analysis on carbon cost. Given that the data come from countries with different economies and energy markets, the LCOEs calculated span a large range of values. It shows that with a 5% discount rate nuclear energy is the most competitive option, while at 10% it remains the most competitive option only in OECD Asia and OECD North America.[4]

Some work includes uncertainty in NPV calculations adopting a probabilistic approach (Roques *et al.*, 2006b). Here the authors calculate the NPV for a new coal, CCGT and nuclear power plants in the United Kingdom market assigning a normal probability distribution to each technical and economical input, including electricity and carbon prices. The resulting NPV itself is given in probabilistic terms, i.e., in terms of a mean and a variance. In this way price volatility is included in the variance of the NPV. In terms of carbon pricing, their study considers a carbon tax normally distributed with mean value of GBP 40 per tonne of CO_2 and standard deviation of GBP 10. Again, with a 5% discount rate nuclear has the highest NPV. With a 10% rate instead, CCGT is the most competitive form of electricity generation.

---

3.   There is a vast literature on how choosing the right discount rate (see Brealey, Allen and Myers, 2006). In general the weighted average cost of capital (WACC) is considered. WACC is given by the weighted sum of the "cost of equity" (the expected rate due to the shareholders) and the "cost of debt" (the cost of the monies borrowed from debt-holders), with the relative amounts of equity and debt as weights. For more details on the discount rate for investment in generating electricity see Chapter 8 of IEA/NEA, 2010.

4.   For a comprehensive list of other studies on the LCOE the reader can look at Chapter 11 of IEA/NEA, 2010.

3.4 Real option analysis

Real option analysis incorporates price uncertainties in investment decisions applying the technique of the call or put option valuation developed in finance (Dixit and Pindyck, 1994). A real option entails the possibility, but not the obligation, for an investor to undertake a business decision such as making, suspending or abandoning an investment. Real option analysis takes into account the value of this management flexibility and incorporates it into the cash flow analysis, that is into the calculation of the NPV. Real option analysis is a valuable method in the presence of future uncertainties when the investor may contemplate different investing opportunities depending on future values of key variables.

Real option analysis works with stochastic variables, i.e., variables whose values are random but whose statistical distribution is known. There exists no option value under perfect foresight. The prices of energy, fuel and carbon may be approximated as stochastic variables. They have high volatility, and it is hard to predict their value day by day, nonetheless over a long timescale, it is possible to identify the main trends and to make use of statistical analysis. In the case of a single stochastic variable an analytic solution can often be found, otherwise one has to rely on numerical solutions and Monte Carlo simulations, where solutions are generated by repeated sampling over random values of stochastic variables, are extensively used.

In analysing nuclear power investment under carbon pricing, real option analysis has been used to incorporate different strategic decisions (or options). One of these is the waiting option or the possibility that the investor can delay the investment. Basic NPV analysis as described above only addresses the question whether it is economically more convenient to invest now or not to invest at all; however, very often an investor wants to wait and see the trends of some key variables before making a decision. For example, in the presence of uncertainties on the future price of carbon, an investor may prefer waiting to see where the carbon price goes before starting the investment. If the price of carbon will turn out to be low, he/she may invest in gas or coal, if it will be high, he/she may abandon the idea to invest in carbon emitting power plants and consider investing in nuclear. An investment with a waiting option has a higher NPV than the same investment without it. The difference between the two NPVs is called *option value* and it is the additional value that comes from the possibility to wait.

A parameter that is often calculated is the *investment threshold*, i.e., the difference between the discounted total revenue of the project minus the discounted investment costs, which determines whether a project should be pursued (Dixit and Pindyck, 1994). If the waiting value is not considered, the investment threshold is zero, since as soon the discounted total revenue of the project equals the discounted investment costs the investment should be pursued. On the other hand, with a positive option value, it may be more convenient to wait even if investing immediately already generates enough revenues to balance the cost. This happens when investing in the future yields revenues that are higher than those generated by an immediate investment. Rothwell first calculated the threshold value for a new investment in a nuclear power plant in the presence of price volatility (Rothwell, 2006).

The IEA has performed some quantitative evaluations of the impact of energy market uncertainty and climate change policy uncertainty (IEA, 2007). This is mainly focused on coal and gas, but analyses nuclear as well. It reports the threshold value for investing in a new coal, CCGT and nuclear power plant. Gas and carbon prices are modelled as stochastic variables and electricity price is determined by marginal costs. The threshold values are calculated under several scenarios with different electricity prices and sources of uncertainty. Scenarios with and without uncertainties on carbon and fuel prices are also considered. When both fuel and carbon uncertainties are taken into account, nuclear investment appears always to be the most risky.

As discussed in Chapter 2, another form of real option is the suspension option, the operating flexibility to interrupt production if revenues are lower than variable costs. The suspension option benefits gas more (and to a lesser extent coal) than nuclear. This is due to two reasons. First nuclear has low marginal cost, thus the option to suspend production due to prices lower than its marginal cost will only happen rarely, and it is not very valuable. Gas instead has high marginal cost, and having prices lower than average cost is not a rare event, thus the ability to suspend production is a very attractive opportunity. Second, capital costs are very high for nuclear and relatively low for gas. Thus in the pessimistic scenario of very low electricity prices, the investor on nuclear power plant risks comparatively high losses, with or without suspension option. The investor on gas instead may use the suspension option and in the worst case shut down the facility, losing only the low initial investment.

Roques *et al.* (2006b) calculate the NPV of investing in a new CCGT and nuclear power plants in the presence of carbon pricing where the CCGT facility has a suspension option. They assume that the plant can be switched on and off at no costs. Their calculations show that the suspension option increases the NPV of the investment, because only positive profits are taken into account, and it increases the competitiveness of gas making it a more attractive than nuclear. Also this study includes a real option analysis with a suspension option under both a carbon trading system and a carbon tax. Three different price scenarios based on the data from the EU ETS will be considered: a base case scenario, a high price scenario and a low price scenario. Of course, the suspension option is most valuable in a low price scenario (see Chapter 5).

A third kind of real option is analysed in Roques *et al.* (2006a) who estimate the option value of keeping open the choice between nuclear and gas in the presence of carbon price. After comparing the NPV of building a new nuclear power plant versus a new CCGT power plant without any option, the authors consider a hypothetical investment consisting of 5 power stations over 20 years where the manager invests in a new power plant every 5 years. This modular approach allows reacting flexibly to developments in fuel and carbon markets, whose prices are modelled as stochastic variables. At a 10% discount rate, the final outcome depends on the correlations between gas and electricity prices. Without price correlation, nuclear is competitive with gas. With correlation gas is preferred even in the presence of carbon pricing as the correlation between gas and electricity prices reduces the risk to invest in gas.

3.5   Portfolio analysis

In finance, portfolio theory aims at finding the portfolio with the highest return given a certain level of risk or, alternatively, the portfolio with the lowest risk given a certain level of profitability. It is based on the concept of diversification, attempting to build portfolios whose collective risk at a given level of profitability is lower than the risk of any single asset at the same average profitability. Even if an asset is very risky, it may be still convenient to invest in it, because its correlation with the other assets may reduce the total risk of the portfolio. The portfolio theory is based on a mean-variance analysis and in general it assumes that assets are normally distributed where the risk is the standard deviation. In a liberalised energy market, companies invest in portfolios of different power plants in order to reduce risks.

Roques, Newburry and Nuttal (2008) assess the impact of fuel, electricity and carbon price and their degree of correlation for an optimal plant portfolio for the United Kingdom market. Monte Carlo simulations are used to calculate the mean and variance of the NPV of investing in a new power plant. A discount rate of 10% and a carbon price normally distributed with a mean of GBP 49 per tonne of CO_2 and standard deviation of GBP 10 per tonne of CO_2 are considered. If electricity, fuel and gas prices move independently, an optimal portfolio would contain a mix of gas, coal and nuclear power plants. However, if gas and electricity prices are highly correlated, an optimal portfolio would contain mostly CCGT plants. The study also examines a portfolio where investors can secure a long-term fixed price power purchase agreement. This would reduce the risk on investing in nuclear and an optimal portfolio would have a balanced mix of CCGT and nuclear power plants.

Portfolio analyses are also considered in Roques *et al.* (2006b) and Green (2008). Roques *et al.* make a comparison of risk-return profiles of different portfolios of power plants. The result of the study is that introducing nuclear plants in a portfolio reduces the likelihood of making large losses due to gas and carbon price uncertainties. Green calculates the share of coal, gas and nuclear in an optimal portfolio and proves that an optimal portfolio would contain a higher proportion of nuclear power plant with a carbon tax than with a carbon trading system.

3.6   EU ETS analysis

There are many studies and analysis on carbon emissions trading system and in particular on the EU ETS, and it would be too long and beyond the goal of this report to try to make even just a synthetic summary of the main works. Here are only reported a few studies that are directly relevant to this study. As presented in Chapter 2, the EU ETS includes more than 11 000 installations representing approximately 40% of EU CO_2 emissions. During Phase I (2005-2007) all allocation was costless. Phase II (2008-2013) is currently still underway with prices hovering between EUR 12 and EUR 18 per tonne of CO_2. In Phase III (2013-2020) almost all allocations will be auctioned.

A first aspect that needs to be considered when analysing the European carbon market is the effect of free allocations. The evidence of the first three-year phase of the EU ETS points towards substantial *gains* for carbon emitters, i.e. fossil-fuel-based power generators, due to the costless attribution of carbon permits. Auctioning will change this dramatically. While carbon-free producers will continue to benefit from higher infra-marginal rents due to higher electricity prices but no carbon costs, carbon-intensive producers will face losses due to the allowances they have to pay.

The work of Burtraw and Palmer (2007) analyses how to compensate the costs posed on the electricity sector (on producers and consumers) by carbon emissions trading using a detailed simulation model of the US electricity sector. Free allocation of all the allowances is not an efficient way to compensate the several actors in the electricity market. They found that local authorities are more efficient in managing allocations because they have access to facility-level information and that full compensation for carbon trading may be achieved by allocating freely only 39% of the emissions allowances.

For a comprehensive presentation and analysis of the working of the EU ETS during Phase I, see Ellerman, Convey and de Perthuis (2010) who provide a good synthesis of current research. They also highlight that during Phase I (2005-2007) the EU ETS achieved a reduction of 2-5% of CO_2 emissions and fundamental changes in the mentality of market operators and relevant institutions that have integrated carbon prices into production and investment decisions. CO_2 pricing has not affected the competitiveness of the industry and the costs for reducing the emissions have been relatively small.

A paper focusing on price formation in the EU ETS is Keppler and Mansanet-Bataller (2010), which studies the causal links between daily carbon, electricity and gas price as well as weather data in the EU ETS. With the help of Granger causality tests the authors show that forward electricity and carbon prices depend in the short run mainly on weather and gas prices, with an additional causal impact provided through the spot market and the market power of operators. An interesting change takes place from Phase I to Phase II where electricity prices begin to drive carbon prices. In reality carbon prices thus capture the residual monopoly rent of electricity producers that is generated in the electricity market rather than being determined autonomously. The results show that standard assumptions about causality (the cost of carbon abatement will determine carbon prices which will set electricity prices) have to be treated with great caution and that only empirical analysis can ultimately determine the impact on different technologies.

3.7 Conclusion

Nuclear energy does not emit any CO_2 during production and carbon pricing, in the form of a tax or of an ETS with auctioning, will increase its competitiveness. However, there is no general consensus on *how much* nuclear will benefit from it and what is the best carbon regime it should hope for. It is important to keep in mind that the EU ETS, the only carbon market in the world, only started in 2005 with a first trial period of three years. Thus carbon market is thus still very young.

In the studies that estimate nuclear investment under a carbon price regime, four different approaches have been recognised. The profit analysis compares the profits of incumbent power plants under different carbon price regimes. This approach only considers performances of operating power plants and does not analyse making new investment. Basic cash flow analysis using standard NPV and LCOE calculations does not fully account for price volatility and is thus more suitable for assessing investment in a regulated market and studying the effects of a carbon tax. Real option analysis tries to estimate how prices volatilities effect the investment using the real option method developed in finance. Portfolio analysis considers investing in a portfolio of different plants and calculates how the carbon regime changes the percentage of nuclear power plants in an optimal portfolio. Finally, a number of studies are examining the fundamental working of the emissions markets with sometimes surprising results.

The first competitor of nuclear under a carbon regime is gas. Current research, however, does not unequivocally answer the question *at what level* a carbon price will make an investor prefer nuclear to gas. The results of the studies mentioned above make different assumptions about carbon, electricity and fuel prices, as well as their volatilities and correlation. Indeed what emerges from the literature is that these prices, their volatilities and correlations play a central role in a liberalised market. For instance, assumptions about correlations between electricity and gas prices play a central role in the investment analysis.

Another key point is the correlation between carbon and electricity prices. If carbon prices are uncorrelated with electricity prices, a carbon trading system would increase the competitiveness of nuclear more than a carbon tax, because the carbon volatility would increase the uncertainties on revenue for gas. On the other hand, if there is correlation and the cost of the carbon entirely passes through the electricity cost, carbon cost would be completely recovered by carbon emitting power plants, and carbon pricing would not affect much the competitiveness of gas versus nuclear. The following chapters will go some way to clarify the empirical, historical relationships and their implications for competitiveness. It is obvious that this will not answer all questions but it will allow future research to advance the issue further.

Chapter 4

Carbon pricing: the competitiveness of nuclear power in LCOE analysis

The interest of carbon pricing for nuclear energy becomes immediately obvious when one considers the CO_2 emissions of coal- and gas-fired power generation in comparison with other generation sources. It is immediately intuitive that the competitiveness of nuclear against coal and gas improves as soon as a price on carbon emissions, whether in the context of an emissions trading system or through a carbon tax, is imposed.

The present study is primarily concerned with analysing the impact of carbon pricing on the profitability concerns of a private investor in an environment of liberalised markets with daily variations in the price of electricity and carbon. It is nevertheless instructive to consider the impact of carbon pricing on the competitiveness of nuclear energy in levelised cost analysis. As indicated above, LCOE analysis develops a notion of social resource cost leading to socially optimal choices with stable prices and costs rather to privately optimal choices in an environment characterised by changing prices and costs under the peculiar pricing mechanisms of electricity markets. Under the assumption of stable prices and costs and using the LCOE methodology, nuclear energy is highly competitive at even modest carbon prices. Later chapters using a different methodology will show that the ability of natural gas to adapt to different price environments in liberalised markets with volatile prices compensates to some extent for its carbon handicap from the point of view of private investors.

This can lead to the divergence of privately optimal choices and socially optimal choices. In certain cost ranges, private investors seeking to maximise their profits in the face of volatile prices in liberalised electricity markets will opt for CCGT, while the socially optimal, cost-minimising choice would have been a nuclear power plant. The latter, however, would only be attractive to investors in an environment where power prices are stable and predictable. It is thus important in these cost comparisons to specify precisely the methodology and the implicitly assumed regulatory environment.

The LCOE analysis embodied in Figures 4.2, 4.3 and 4.4 assumes a 7% real interest rate and the technical specifications of the European median case in the study *Projected Costs of Generating Electricity*. It shows that in Europe nuclear energy is competitive against coal at a carbon price of around EUR 15 (USD 22) per tonne of CO_2, which corresponds closely to the current market price on the EU Emissions Trading System.[1] Clearly, the competitiveness of coal deteriorates very quickly with an increasing carbon price. In an LCOE methodology, as is appropriate for markets in which regulators interested in the minimisation of generating costs set the prices, gas is never truly competitive with nuclear energy at a 7% discount rate in baseload power generation.

---

1. Since most of this study is based on data from European power markets as well as the EU ETS, values for this and the following figures in this chapter are indicated in Euros, although the results are drawn from the Projected Costs study, which was entirely denominated in US dollars.

Figure 4.1: Direct and indirect CO_2 emissions of different power generation technologies

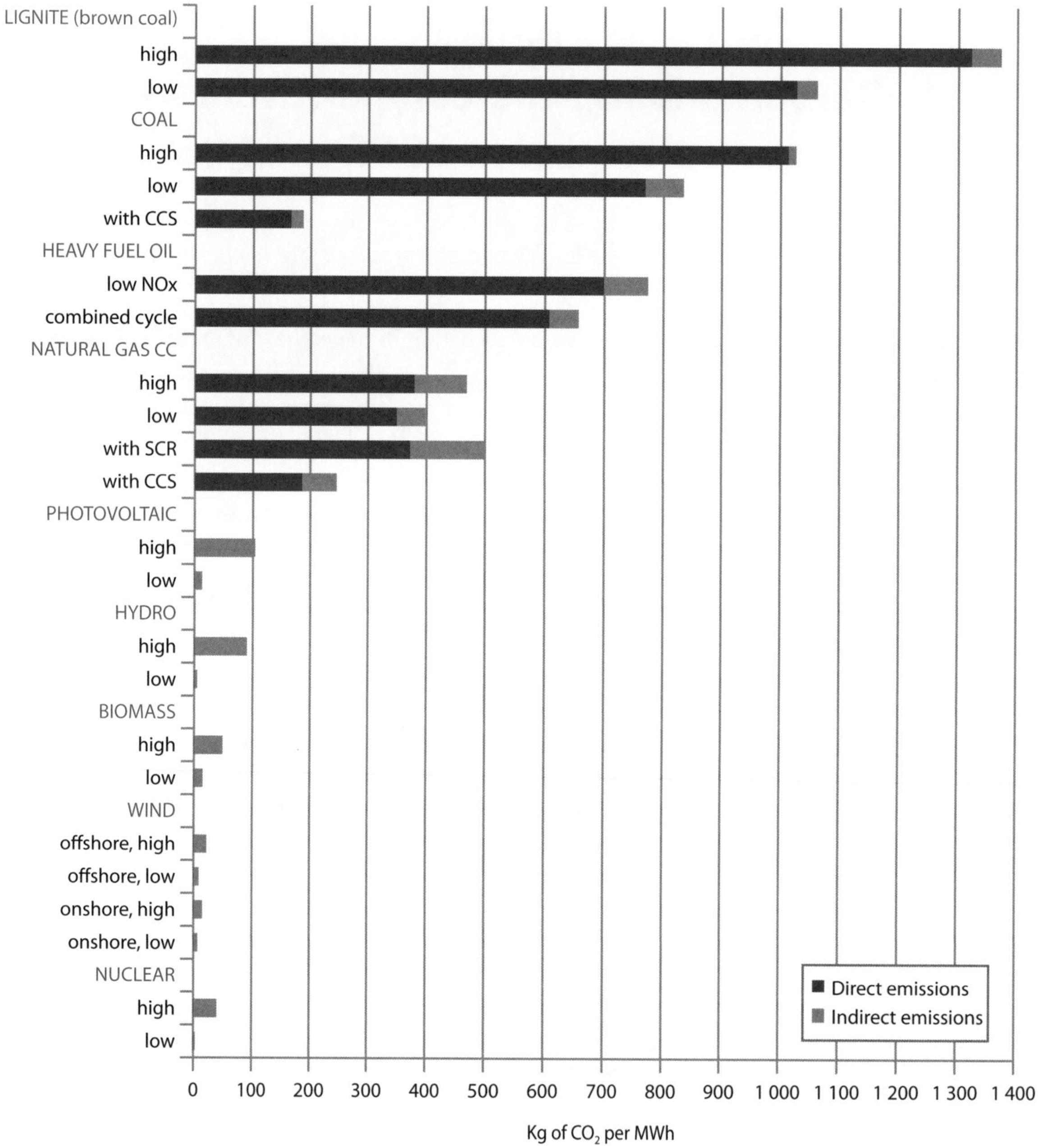

Source: IPCC, 2007, 4.3.4.1 Electricity.

Figure 4.2: Carbon pricing and the competitiveness of nuclear energy in OECD Europe
LCOE of different power generation technologies at a 7% discount rate

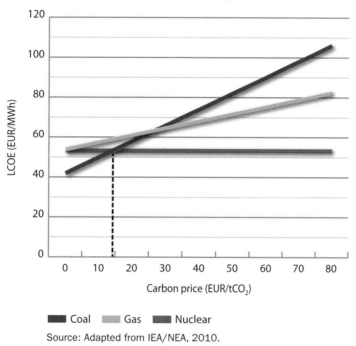

Source: Adapted from IEA/NEA, 2010.

Figure 4.3: Carbon pricing and the competitiveness of nuclear energy in OECD Asia-Pacific
LCOE of different power generation technologies at a 7% discount rate

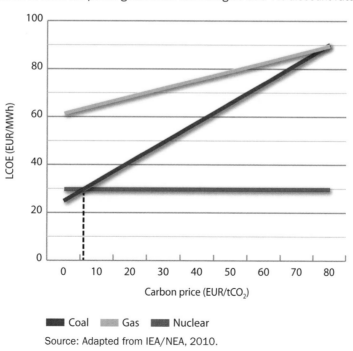

Source: Adapted from IEA/NEA, 2010.

The situation is quite different in OECD Asia-Pacific and OECD North America. In the Asia-Pacific region, nuclear energy thus becomes already competitive at around EUR 6 (USD 9) per tonne of CO_2 (see Figure 4.3). The figure also shows the impact of relatively higher gas prices in this region that make natural gas uncompetitive as a baseload technology in most cases.

In OECD North America, nuclear energy becomes competitive against coal at a carbon price of around EUR 17 (USD 24) in an LCOE methodology, which closely mirrors the European situation (Figure 4.4). At very low-carbon prices and the assumptions of the LCOE study, also gas-based power generation is competitive against nuclear. Recent developments in North American gas markets might even increase this advantage.

However, it should be kept in mind that contributions to the Projected Costs study from OECD Europe and OECD North America provided data for first-of-a-kind Generation 3+ nuclear reactors, of which the first few pilot plants are currently being built. One may thus assume that unit costs may come down considerably in the future once economics of scale and learning effects increase efficiency and drive down costs. To some extent the data from OECD Asia-Pacific validate this hypothesis, since the considerably lower unit costs provided for this region pertain to Generation 2 reactors, whose unit costs already benefit from the economies of scale of dozens of existing reactors and several decades worth of learning. The differences in three OECD regions thus relate not only to different industrial manufacturing cultures but also different environments for regulation and public acceptance of nuclear plants.

Figure 4.4: Carbon pricing and the competitiveness of nuclear energy in OECD North America
LCOE of different power generation technologies at a 7% discount rate

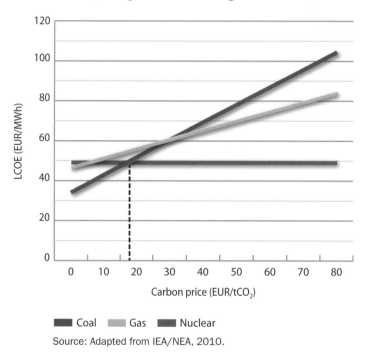

Source: Adapted from IEA/NEA, 2010.

4.1   Paying or not paying for CO_2 emissions?

The question of "paying or not paying for CO_2 emissions?" might seem odd in a publication on the impact of carbon pricing on the competitiveness of nuclear energy. To some extent it is. The previous figures, as well as much other theoretical analysis, clearly assume that fossil-fuel-based power producers *pay* for their CO_2 emissions. However, when the problem is approached from the empirical side, the question is anything but odd. In fact, in the only significant carbon pricing system currently operating, the EU ETS, emitters do *not pay* for their emissions. Instead the carbon emitters included in the system, including power producers which make up around 50% of emissions in the EU ETS, receive their emissions permits for free.

The number of permits corresponds to historic emissions minus a small percentage reduction reflecting the system's overall reduction objective, a practice called "grandfathering" since it extends the historic allocation of emissions rights in favour of large-scale emitters. "Grandfathering" is indeed the practice pursued for the vast majority of permits during the first two phases of the EU ETS, 2005-07 and 2008-12. In function of their actual emissions, emitters can then decide to either reduce their own emissions by the required amount or buy the difference on the carbon market. In either case, emitters would undergo at most the marginal costs of adjustments rather than the full costs of their emissions.

However, the peculiar pricing mechanisms of electricity market and the fact that carbon-intensive producers include the opportunity cost rather than the actual cost into power prices resulted in substantial windfall profits (also referred to as carbon rents) for *all* electricity producers including those based on carbon-intensive coal and gas since the introduction of the EU ETS in 2005 (see Box 4.1). While all producers gained as a result of higher electricity prices as long as permits are given out for free, low-carbon and carbon-intensive producers react in completely different ways to a switch from free allocation to paid-for allocation. Paying for permits typically takes the form of organising auctions of emissions permits at which fossil-fuel-based producers buy the permits they need from their respective governments. The impact of the switch towards auctioning varies dramatically between low-carbon producers and fossil-fuel-based producers:

- *Low-carbon producers* such as nuclear or renewables profit from a carbon trading scheme regardless of whether permits are grandfathered or auctioned. In both cases, they profit from higher prices for electricity while their costs stay the same. The fact that prices rise also when permits are allocated for free is due to the principle of opportunity cost and has been amply verified in European electricity markets since introduction of the EU ETS.

- *Carbon-intensive producers* such as coal- and gas-based power producers (only small amounts of oil are used for power generation in Europe) also profit from higher prices, in particular as long as permits are allocated for free. As soon as permits are auctioned off, however, their costs will rise and their windfall gains will be reduced. Their net position will depend on their carbon intensity. While coal-based producers will lose in comparison to a situation without carbon pricing, gas-based producers are still likely to experience a small gain even with carbon pricing when permits are auctioned off.

Box 4.1: Understanding "opportunity costs" and the windfall profits of carbon-intensive producers with a free allocation of carbon permits

The principle of opportunity cost ensures that utilities will include the market price of a carbon permit into the price of their electricity even if they have received the permit for free. This has generated much questioning and criticism. The process, however, is natural from an economic point of view. Imagine a coal-based utility that emits roughly one tonne of CO_2 per MWh of electricity. Imagine further that its cost of production (net of CO_2) are EUR 40 per MWh, the price of a permit is EUR 20 on the EU ETS, that the coal-based power producer has received his/her permits for free and that both the electricity market and the carbon market are competitive. The question is now whether the utility will sell its output at EUR 40 per MWh (its true cost) or at EUR 60 per MWh (its opportunity cost). The correct answer is EUR 60 per MWh.

Why is that so, given that the permit was received for free? In order to understand the principle of opportunity cost, one must think of the profit situation of the utility if it would *not* produce electricity (an "opportunity" it would forego by producing). In this case it would save EUR 40 per MWh on production costs, sell the permit on the EU ETS and make a profit of EUR 20. Thus asking for EUR 60 (and using the permit in the process) is the minimum amount necessary to induce the utility to produce. Only the EUR 60 price allows earning an equivalent EUR 20 through the production of electricity. Producing or not producing, once the utility has received a valuable carbon permit for free, it is unequivocally better off than before.

The amount of the price of the permit that is passed on to electricity consumers is referred to as "pass-through". It can usually be assumed to be 100%. It is important to understand that due to the principle of opportunity cost, a 100% pass-through would prevail in particular under perfect competition with perfectly elastic, horizontal demand curves. There are configurations for inelastic demand curves, i.e., demand curves allowing for a degree of monopoly power, where pass-through rates may diverge from 100%. It can be shown that in the case of linear, downward-sloping demand curves, the theoretically optimal values for pass-through may be somewhat lower than 100% and that in the case of isoelastic demand curves theoretically optimal values may be somewhat higher than 100% (see Keppler and Cruciani, 2010). In the absence of specific information about the shape of demand curves, however, the direction of the divergence cannot be established and working with a pass-through rate of 100% is clearly the appropriate assumption for empirical analysis.

While the European Commission has announced a switch to full auctioning of all permits in the electricity sector with the beginning of Phase III in 2013, the existence of windfall profits for both low-carbon and high-carbon producers has lead to much public criticism. Free allocation during an initial introductory period was probably necessary to include carbon-intensive power producers in the economic and political coalition at the European level that supported the creation of the EU ETS. Nevertheless it is fair to say that the introduction of the EU ETS benefited European power utilities which saw their market capitalisations increase rapidly after 2005. This dynamic has not only stopped but partly been reversed in 2008, when the announcement of the new carbon market realities after 2013 massively affected the utility sector over and beyond other factors. The impact of the financial and economic crisis was thus far more intense that in other sectors. In addition to the opening of European electricity markets, the added profits contributed to the intense merger activity between European utilities during the past five years.

The dynamics are magnified if one compares in Figure 4.5 below the almost completely coal-based and thus highly carbon-intensive UK power producer Drax Plc. with nuclear-based Électricité de France (EDF), and the German utilities E.ON and RWE with a mixed portfolio of generation assets. While national circumstances certainly played a role in all cases, it is nevertheless easy to see how carbon-intensive Drax outperforms other European utilities, in the early days of the EU ETS. Stagnation and decline set in once future carbon liabilities are taken into account by investors. The figures would have been even more impressive if 2005 had been chosen as base year but neither Drax nor EDF were traded at that moment in their present corporate structure.

Figure 4.5: Market capitalisation of Drax, EDF, E.ON and RWE since 2006

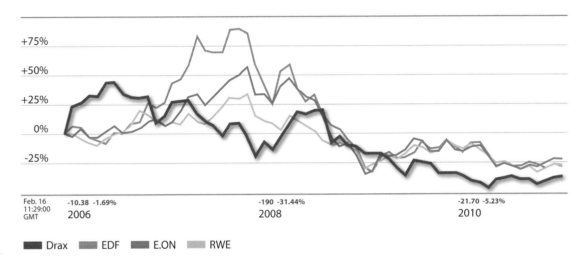

Source: *Financial Times,* http://markets.ft.com/markets/interactiveChart.asp.

The key issue is that for carbon-intensive fossil-fuel technologies free permit allocation through "grandfathering" is substantially more profitable than auctioning. Table 4.1 below shows estimates of the rents and the additional costs that European power producers either obtained in Phase I with grandfathering or will have to pay during Phase III when generalised auctioning have been introduced for European electricity producers (for purposes of comparison, the calculations assume that the auction price will correspond to the average Phase I price). It is obvious that from 2013 onwards auctioning will impose a significant additional cost on fossil-fuel-based power producers and thus enhance the competitiveness of nuclear power. At a carbon price of EUR 12, the difference between the two modes of allocation is in the order of EUR 11 billion for coal-fired power production and in the order of EUR 2.5 billion for gas-fired power production.

Table 4.1: The different impacts of free allocation and auctioning*

Annual rents or losses due to carbon pricing, million Euros and EUR $12/tCO_2$ average price

Technology	TWh	Rents before EU ETS	Rents with EU ETS		Additional rents due to EU ETS		Difference FA-AU
			Free allocation	Auctioning	Free allocation	Auctioning	
Nuclear	998	16 325	21 791	21 791	5 466	5 466	0
Coal	1 001	11 137	17 657	6 848	6 520	- 4 289	10 809
Gas	664	3 572	7 141	4 740	3 569	1 168	2 401

Source: Adapted from Keppler and Cruciani, 2010, p. 4289.

* In order to facilitate the understanding on this comparison we recall that all technologies earn "rents" even without the introduction of a carbon trading system. These rents are constituted by the difference between a technology's variable cost and the market price that is set exclusively by the technology with the highest variable cost, which depending on varying market conditions may be oil, gas combustion turbines or coal (the calculations for gas use the variable cost for combined-cycle turbines). Thus gas-fired power generation can still profit from higher price with auctioning despite higher variable costs.

Clearly, coal-based producers fare very badly in carbon regimes in which emitters actually pay for their emissions, such as an auctioning or a carbon tax. While this poses stark financial and commercial issues for these producers, it is also clear that the original idea behind carbon pricing was precisely driven by the desire to enhance the competitiveness of low-carbon generating technologies such as nuclear and renewables. Paying for carbon emissions is the most efficient and absolutely indispensable solution to internalise the negative externality of climate change. In this sense, carbon emissions trading with a free allocation of emissions permits was always an exception, a temporary measure to ease the transition from a historic state of costless carbon emissions to a new state where the private costs of emitting carbon reflect the social costs.

Chapter 5
Profit analysis

The present study analyses the competitiveness of nuclear power in the presence of carbon pricing with two different methodologies. The first of these methodologies ("profit analysis") focuses on the *short-term* impact of carbon pricing under the EU ETS and an equivalent carbon tax. It thus concentrates on the respective profits that operators made during the 2005-10 period due to the introduction of the EU ETS. In has the great advantage of being able to work with real historical price and cost data and thus provides quite a realistic picture of events.

By definition, its limitation is the fact that production costs are confined to variable costs. Profit analysis does not take into account the investment costs of electricity generation and thus has nothing to say on the actual or potential investment decisions of operators, a topic that is dealt with in the investment analysis in Chapter 6, which will concentrate on the *long-term* impacts of different forms of carbon pricing in particular on new investments. As has been pointed out above (Chapter 2), both methodologies will work with a combination of data from IEA/NEA (2010) and daily price and cost data from European energy and carbon markets between July 2005 and June 2010.

5.1 European energy and carbon prices from 2005 to 2010

Since the creation of a spot market for CO_2 permits under the EU ETS in June 2005 until mid-2010, all markets have undergone enormous upheaval in conjunction with rapid global growth until the summer of 2008 and the economic and financial crisis that followed. European energy markets were no exception and if anything exacerbated by these violent swings due to a number of sectoral and regional issues:

- Europe's electricity markets were heavily impacted by the introduction of the EU ETS due to the phenomenon of "pass-through", the inclusion of carbon prices in electricity prices (see Chapter 4). Due to the fact that electricity is not storable and investments have long time lags, markets for electricity also react stronger than other markets to shifts in demand.

- In the EU ETS carbon market, exaggerated expectations and speculation led to prices as high as EUR 30 per tonne of CO_2; these prices collapsed essentially to zero when it became clear that carbon permits had been over-allocated during the 2005-07 Phase. During Phase II, 2008-12, prices so far are trading between EUR 12 and EUR 18, with EUR 15 being widely seen as a politically acceptable target price.

- The European gas market was rattled by security of supply fears during 2006 due to a combination of declining United Kingdom production, increasing Russian domestic demand and tensions between Russia, which provides 25% of European gas, and Ukraine, a major transit country.

- The essentially global coal market, which had traded for decades below or at around EUR 50 per tonne of hard coal, saw huge rises in 2008 with prices above EUR 150 per tonne due to substantial increases in Chinese demand, where economic growth rates of 10% implied additions of coal-based power capacity of up 70 GW (the total installed capacity of the United Kingdom) per year. While the financial and economic crisis brought down coal prices almost to EUR 50, they have recently risen again to around EUR 90 per tonne.

Figure 5.1 provides a plot of daily price data in these four markets.

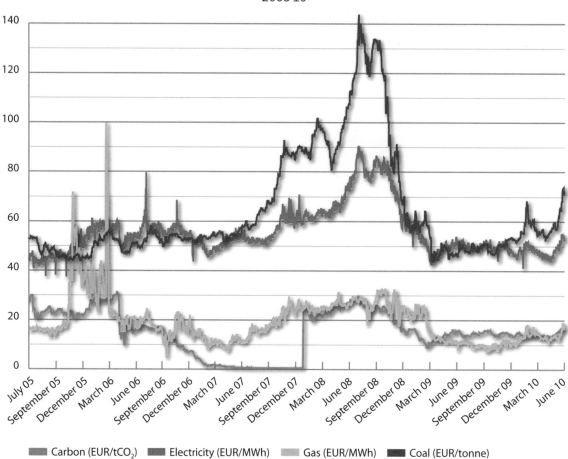

Figure 5.1: European prices for electricity, carbon, gas and coal
2005-10

Working with data from such a volatile period has advantages and drawbacks for the modeller who is looking to identify stable relationships between the different variables. The drawback is that any stable long-term relationships that are bound to emerge through the profit-maximising behaviour of market participants in a calmer environment struggle to become identifiable through the short-term noise generated by exogenous events. Constantly displaced equilibrium relationships are thus difficult to identify. The advantage is, of course, that such a wide range of variations might just be a realistic rendering of the actual workings of the market. This is an issue, in particular, in the investment analysis when the evolution of market prices has to be predicted for the lifetime of the investments, i.e. up to 60 years. Basing these predictions on data that reflect strongly differing situations increases the robustness and realism of the modelling results in the absence of any identifiable central tendency for trends and correlations.

5.2 The profitability of different power generation options in the presence of carbon pricing

The objective of the profitability analysis presented in this chapter is threefold. The first objective is to assess how carbon pricing in the EU ETS has impacted the level profits from electricity production. The study thus calculates in an empirical *ex post* analysis the average per MWh profits for electricity produced by nuclear power plants, coal-fired power plants and combined-cycle gas turbines from 2005 to 2010. It goes on to show the enormous difference in profitability between a free allocation of carbon permits and true carbon pricing with payment for allowances, say through an auctioning scheme. Ideally, the analysis would also have compared average profits during the 2005-10 period with the profitability of earlier periods to identify the precise additional impact of carbon pricing. Establishing a reliable counter-factual, however, is impossible given that not only carbon markets but to a large extent also liberalised electricity markets did not exist before 2005.

The second objective of the study is to relate the average profitability of nuclear, coal and gas to the volatility of their returns both under carbon trading in the EU ETS as well as under an equivalent carbon tax. It is often advanced that nuclear energy has more to gain under a carbon tax than under carbon pricing because of the stability of the carbon price signal. This argument, however, forgets that what is decisive for investors (other than the absolute height of average profits) is not the stability of the carbon price signal but the stability of the profit flow. If, for instance, limited correlation between electricity and carbon prices in a carbon trading system makes profits for coal- and gas-based electricity producers more volatile than with a carbon tax, then the relative competitiveness of nuclear energy is *enhanced* through carbon trading. As the results below show, the differences are not enormous, and should be read in a manner that nuclear energy has nothing to fear from carbon trading even if this means more volatile carbon prices. Its competitors will have relatively more to lose.

The third purpose is to determine the monetary value of what is referred to throughout this study as the "suspension option", the ability of technologies with a comparatively low fixed-cost-to-variable-cost ratio, such as gas, to leave the market when electricity prices are low. This is also frequently referred in discussions as the "flexibility" advantage of low fixed cost technologies. In technical terms, the value of the "real option" of a gas-based power producer to leave the market when prices are low is greater than the corresponding value for a nuclear-based producer who will have to undergo passively a spell of low prices (see Chapter 2). The numerical analysis below shows that the value of the "suspension option" is noticeable but less significant than it is sometimes implicitly assumed in expert discussions. In the following the results to the three questions are provided one by one.

Average profits for nuclear, coal and gas under the EU ETS

As mentioned, the analysis in this chapter only applies to existing production facilities and does not assess the relative profitability of investments in new power plants, which is considered in Chapter 6. This means considering only variable costs, which are the costs associated with running an existing power plant. In this first step are thus established the average per unit profits for different power generating technologies. This average is calculated by taking the mean of all daily values during the 2005-10 period. In this first step, daily profits are then calculated the following way:

$$R(t) = P(t) - O\&M - FC(t) - CC(t) \tag{1},$$

where P(t) is the price of electricity, FC(t) is the fuel cost, O&M are the costs for operation and maintenance and CC(t) are carbon costs. This formula assumes fuel carbon pricing; when considering actual historic profits in the EU ETS without carbon pricing carbon prices will not be included. $R(t)$ is a function of time as $P(t)$, $FC(t)$ and $CC(t)$ change daily. Only, the O&M costs are assumed to remain constant at the level determined in IEA/NEA (2010). In practice, O&M costs can, of course, also change over time, but usually such fluctuations are small and not correlated with other variables. Daily carbon costs, $CC(t)$, are calculated by multiplying the carbon content of each fuel, CCT, with the daily carbon price. Daily fuel costs of gas and coal plants instead are calculated by:

$$FC(t) = HC * (1/EFF) * FP(t) \tag{2},$$

where HC is the gross calorific value (heat content) of the respective fuel, EFF is the technical conversion efficiency of converting fossil fuels into electricity and $FP(t)$ the daily fuel price. Fuel costs for nuclear energy are assumed to be fixed. All values are normalised to the dimension of one MWh of electricity. In a second step, average daily profits are calculated. This yields for coal and gas:

$$R = \text{Average } [R(t) = P(t) - O\&M_{COAL, GAS} - FC(t) * P_{COAL, GAS}(t) - CCT_{COAL, GAS} * P_{CO2}(t)] \tag{2a},$$

and for nuclear energy

$$R = \text{Average } [R(t) = P(t) - O\&M_{NUCLEAR} - FC_{NUCLEAR}] \tag{2b}.$$

The data for the stable carbon content, heat content, conversion efficiency, O&M costs and nuclear fuel costs are again taken from IEA/NEA (2010), while daily price data come from the sources presented in Chapter 2.

On the basis of equations (2a) and (2b), three values have been calculated for electricity produced on the basis of nuclear, coal or gas:

1. The real historic average profits per MWh that have been generated by the different technologies during the period of analysis, July 2005-May 2010. Since CO_2 emission permits were allocated gratuitously during this period, these profits correspond to a case of "free allocation".

2. The average profits that would have been generated if CO_2 permits would have had to be paid for either on the EU ETS market or through an auction mechanism. Since an auction mechanism is to be introduced in 2013, this case is referred to as post-2012 auctioning. It remains, however, based on prices observed from 2005 to 2010.

3. The average profits that would have been generated if a carbon tax had been levied on the CO_2 emissions of coal- and gas-based power producers. To ensure comparability, the carbon tax is assumed to be equal to the average of the observed carbon prices during the 2005-10 period. Electricity prices are assumed to be unchanged from the two earlier cases.[1]

---

1. This last assumption is less innocent than it may appear. Due to the linkages between carbon and electricity prices, one can expect changes in electricity prices once volatile carbon prices are substituted by a stable carbon tax. Ultimately the question is one of the causality between electricity and carbon prices. If electricity prices cause carbon prices (which is likely to be the case in the short run) then the assumption of unchanged electricity prices is the correct one. However, if carbon prices cause electricity prices, then clearly the switch from market pricing to a carbon tax would also impact electricity prices. Current evidence on this point is not entirely conclusive but there is some evidence for electricity prices causing carbon prices since 2008 (see Keppler and Mansanet-Bataller [2010] for a detailed discussion of causalities between electricity, carbon and fuel prices).

While quantitative results are provided in Table 5.1 below, Figure 5.2 already enables identification of the key qualitative messages. During the past five years operating an existing nuclear power plant has been an extremely profitable affair. Of the three main technologies, it was by far the most profitable one. In truth, anything else would have been surprising given that nuclear energy has the highest share of fixed investment costs that need to be repaid through higher than average operating profits. Nevertheless, nuclear energy was highly profitable during the past five years. In addition, as a carbon-free technology during operations, its profitability will not be affected by a switch from a free allocation of permits to auctioning.

Figure 5.2: Average profits with suspension option

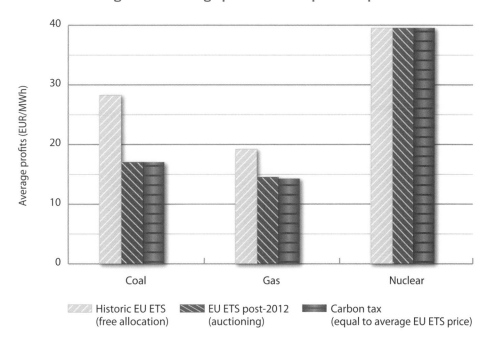

Coal and gas also earned very respectable average operating profits from 2005 to 2010. Given that the fixed-cost-average-cost ratio for coal is somewhat higher than for gas, also its operating profits are somewhat higher. Noteworthy, however, are two particular aspects. First, there is the sharp drop in the operating profits of coal once carbon pricing is introduced. This is a result that will be confirmed by the investment analysis in Chapter 6. Carbon pricing even at the relative modest amount of EUR 14 per tonne of CO_2 (the average price of carbon from 2005 to 2010) has an enormous impact on the competitiveness of coal. Add to this the rising coal price due to continuing Asian demand and it is very unlikely that new coal plants will be competitive against gas and nuclear energy in OECD Europe or any other region with a significant price of carbon.

Second, the difference in average competitiveness between a volatile carbon price and a stable carbon tax (the latter being calculated as the average of the former) is very small. In fact the difference is due only to the slightly different moments at which the "suspension option" is exercised. The far more interesting question of whether the switch from carbon pricing to a carbon tax will have an impact on the volatility of profits will be discussed in the following section.

The relative competitiveness of nuclear energy under carbon trading and carbon taxes

One of the objectives of this study was to test the assumption whether the competitiveness of nuclear energy against other technologies would hold up better under a carbon trading regime or a carbon tax. Commentators often implicitly or explicitly assume that carbon taxes would be the preferred instrument to ensure the competitiveness of nuclear energy. The intuition behind this argument is straightforward: nuclear energy as a high fixed cost technology needs a predictable stream of profits – hence a predictable price of carbon is preferable. This reasoning, however, makes the fallacious assumption that a stable stream of profits for nuclear power, which is indeed a welcome quality, depends on a stable price of carbon. This is, however, not automatically the case. *Ceteris paribus*, in particular electricity prices and the *average* carbon price, nuclear energy as a source of electricity that is carbon free during production remains unaffected by the choice of framework for carbon pricing. It is fossil-fuel-based electricity generation whose stream of profits will be affected by carbon pricing. In question is thus the *relative* profitability of nuclear energy in comparison to its two key competitors which are gas and coal.[2]

The objective is thus to calculate the profitability of coal- and gas-fired generation both under a carbon trading system and under a carbon tax. In order to compare like with like, one must assume that the level of the carbon tax is equal to the average carbon price in the EU ETS during the 2005-10 period. Showing that a tax higher (lower) than the average carbon price improves (diminishes) the relative profitability of nuclear is hardly noteworthy. Decisive in this context is only the impact of a switch from carbon trading to a carbon tax on the *volatility* of the profit-stream for coal and gas. The relative profitability of nuclear energy is thus determined by a comparison of the risk-adjusted profit streams of the three technologies.

In order to assess the difference between a trading system and a carbon tax, the height of the average daily profit and their volatilities are assessed for nuclear, coal and gas, considering two different carbon costs. Averages are again calculated as the mean of daily values during the 2005-10 period. The average profits for the first case, carbon trading under the EU ETS, correspond to equation 2a above:

$$R_{Trade} = \text{Average } [R_{Trade}(t) = P(t) - O\&M_{COAL, GAS} - FCT^*P_{COAL, GAS}(t) - CCT^*P_{CO2}(t)] \qquad (3a).$$

Correspondingly, the average profits for the second case, an equivalent carbon tax, are calculated to the analogue equation:

$$R_{Tax} = \text{Average } [R_{Tax}(t) = P(t) - O\&M_{COAL, GAS} - FCT^*P_{COAL, GAS}(t) - CCT^*T_{CO2}] \qquad (3b).$$

R_{Trade} and R_{Tax} thus correspond to the average per MWh profits for coal and gas, which are based on daily returns from July 2005 to May 2010. The only difference between $R_{Tax}(t)$ and $R_{Trade}(t)$ concerns the carbon cost, where T_{CO2} is equal to the average of $P_{CO2}(t)$. In principle, R_{Trade} and R_{Tax} should be identical given that the level of the tax is calculated by taking the average price in the carbon trading system. They differ slightly for coal and gas, however, due to the existence of the suspension option, which will be exercised slightly more often with the more volatile prices under the carbon trading

---

2. While the reasoning that carbon taxes provide better foresight for investors in nuclear energy does not hold from an economic point of view, it may have some merit in a political dimension. It is not impossible that policy makers may find it easier to commit themselves to a given level of carbon tax rather than to a given average price of a carbon trading system. Since the quantitative objectives of a trading system translate only very imperfectly into a given price level, the impact on the competitiveness of the different technologies is difficult to predict. The key point, however, also in this case is that the absolute profitability of nuclear is not affected as long as electricity prices stay the same. What is affected, is the profitability of coal- and gas-based generation and hence the relative competitiveness of nuclear energy. And the following analysis shows that volatile carbon prices *diminish* the profitability of coal and gas.

system. The returns for nuclear remain unchanged from equation 2b, since they are not affected by carbon pricing:

$$R = \text{Average } [R(t) = P(t) - O\&M_{NUCLEAR} - FC_{NUCLEAR}] \tag{3c}.$$

In a second step, the *volatility* of the two different streams is computed where the standard deviation of returns, σ_R, provides the habitual measure for volatility, where

$$\sigma_R = (\text{Average } [(R(t) - \text{Average } R(t)]^2)^{1/2} \tag{4}.$$

In a third and final step, the risk-adjusted profit streams are compared with the help of the Sharpe ratio. Sharpe ratios are a handy and intuitively appealing measure used in financial economics to compare different profit streams with different risk-reward trade-offs. The Sharpe ratio (SR), called also the reward-to-variability ratio of an asset, is defined as the ratio of the average return (profits) of an asset and its standard deviation:

$$SR = \text{Average } R(t)/ \sigma_R \tag{5}.$$

The Sharpe ratio thus allows comparing different streams profits with idiosyncratic volatilities by providing a risk-adjusted measure of return. It indicates to which extent the return of an asset compensates the investor for his/her risk. A high Sharpe ratio means either high return or low risk, thus investors prefer to invest in assets with high Sharpe ratios. Two different pricing scenarios for three technologies imply computing six different Sharpe ratios (or rather five given that the Sharpe ratios for nuclear energy will be identical for the two carbon pricing regimes since the carbon price is zero in both cases) for comparison.

This procedure answers the key question of how the choice of carbon pricing regime affects the ranking of the different profit streams and the relative competeiveness of nuclear under the profit analysis. Whether a carbon tax will increase or decrease the volatility for the operators of gas- and coal-fired power plants in comparison to carbon trading depends primarily on the correlations between carbon and electricity prices. If correlation is absent, then the volatility of revenues for gas- and coal-fired power plants should be lower with a stable carbon tax than with trading, their Sharpe ratios should increase and thus reduce the relative competitiveness of nuclear. In general, if carbon prices are positively correlated with electricity prices then the volatility of revenues for gas- and coal-fired plants should be higher with a carbon tax, their Sharpe ratio should decrease and improve the relative competitiveness of nuclear.[3] Table 5.1 and Figure 5.3 show the results of the empirical analysis based on real world data.

---

3.   While this reasoning is general and the working with Sharpe ratios is a pertinent and transparent manner to compare different profit streams, a limitation of the present research should also be mentioned. This limitation consists in the fact that day-by-day electricity prices were assumed to remain unchanged when moving from a carbon trading scenario to a carbon tax scenario. It is, in fact, conceivable that daily electricity prices change in function of the switch from a variable carbon price under carbon trading to a stable carbon price under a carbon tax. Depending on the assumptions concerning the short-term causality between electricity and carbon prices one might expect changed electricity prices in function of changed carbon prices (see the discussion on "pass-through" in Chapter 4 for more detail). The question is, of course, of relevance to the correlation between electricity and carbon prices and hence the development of the Sharpe ratio. At the same time, one needs to consider that the carbon tax is equal to the average carbon price under carbon trading and that thus average pass-through remains the same in both scenarios. The above analysis would thus hold precisely if one assumed pass-through based on average values while both carbon prices (under trading) and electricity prices would exhibit uncorrelated short-run variations based on exogenous events. This is not unrealistic assumption, as electricity prices react to the short-run meteorological and technical events and carbon prices to institutional events such as the issuance of Certified Emission Reductions (CER) for projects under the Clean Development Mechanism (CDM) or policy announcements of the European Commission. In the end, the above analysis on this particular point should be seen as a starting point for further research rather than as conclusive.

Table 5.1: Average profits, standard deviation and Sharpe ratio for coal, gas and nuclear

	Historic EU ETS (free allocation)			EU ETS with carbon pricing (auctioning)			Carbon tax (EUR 14.40/tCO$_2$)		
	Coal	Gas	Nuclear	Coal	Gas	Nuclear	Coal	Gas	Nuclear
Average profit (EUR/MWh)	28.24	19.17	39.48	16.99	14.48	39.48	16.98	14.22	39.48
Standard deviation (EUR/MWh)	4.75	8.53	9.77	6.73	8.67	9.77	4.75	7.76	9.77
Sharpe ratio	5.95	2.25	4.04	2.52	1.67	4.04	3.58	1.83	4.04

While the risk-adjusted profit of nuclear energy expressed in the Sharpe ratio remains stable, the risk-adjusted profits of gas and coal change markedly. Quite obviously they change, as already highlighted above, as a function of the allocation mechanism, i.e., whether carbon permits are handed out for free (the left-hand bar for each technology) or allocated against payment, for instance through an auction system (the middle bar). The crucial question in this section is, of course, how the middle bar compares to the right-hand bar, which indicated the risk-adjusted profitability under a carbon tax.[4] One can see that for both coal and gas, risk-adjusted profitability is *lower* in a carbon trading system as long as permits have to be paid for (which is the default assumption in the analysis of carbon pricing). Table 5.1 also shows that the *average* profits (unadjusted for volatility) of gas decrease when switching from a trading system to a carbon tax. This is due to the more frequent use of the suspension option in a carbon trading system.

In other words, a carbon tax would increase the Sharpe ratio of both gas and coal with the effect that nuclear is actually *more* competitive under carbon trading. However, the differences are small for gas (<10%), and middling for coal (<33%). Adding to this the methodological issues highlighted in Footnote 4, the results should probably be formulated in a manner that says "nuclear energy has nothing to fear from carbon trading" rather than saying that "nuclear energy should always push for carbon trading over a carbon tax". Crucial is the absolute level of the carbon price over the long term. Once this is assured, nuclear energy can indulge itself in the rare privilege of being (almost) indifferent to the form in which it is administered.

---

4.   Even though it is less directly relevant to the question of whether a carbon trading system or a carbon tax is more favourable for the competitiveness of nuclear power, an interesting question is posed by the very high Sharpe ratio for coal in the absence of carbon pricing, i.e. under the real historic conditions of the first Phase of the EU ETS. The point is all the more interesting as coal actually improves its competitive position when working with Sharpe ratios and taking volatility of profits into account. The reason is, of course, that the volatility of profits for coal is by far the lowest among the three technologies (its standard deviation is about 50% lower than that of nuclear and about one third lower than that of gas), a fact that is masked in the "EU ETS with carbon pricing" case by a sharp drop in average profits. This is due to the very high correlation of almost 0.9 between coal prices and electricity prices. Due to the high coal prices during the 2005-10 period coal was frequently the marginal fuel setting the electricity price. This means that its higher resource costs were offset by higher revenues, hence the low standard deviation of profits. Gas profited less from this effect and nuclear not at all. The existence of long-term gas supply contracts might be one explanation for the lower correlation of 0.5 between gas and electricity prices.

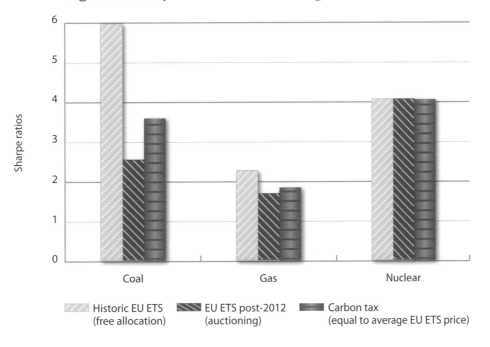

Figure 5.3: Sharpe ratios for carbon trading and a carbon tax

The value of a "suspension option"

The third question pursued in the profit analysis is the monetary value of the "suspension option", the ability for technologies with low fixed-cost-to-variable-cost ratios to profit more than others from the ability to defer production through time when prices are low. On the face of it, this seems a question of relevance for theoretical analysis rather than for practical decision making in either finance or politics, in particular when average profitabilities have already been established independently. This is, however, a partial view.

The precise quantitative determination of the value of a suspension option in the electricity sector on the basis of empirical data is a relevant contribution to current discussions given the degree to which the focus of applied economists has shifted towards the consideration of "real-valued options" in the wake of the seminal contribution by Dixit and Pindyck. The rather modest value of the suspension option, however, goes some way to dispelling the fear that its unavoidable omission in certain methodologies, for instance, in LCOE analysis, introduces a significant bias in favour of high fixed cost, low variable cost technologies such as nuclear.[5] Gas as the technology with the lowest fixed-cost-to-variable-cost ratio among the main technologies of roughly one-to-two does indeed profit

---

5. The quantitative analysis does not include ramp costs, the increase in variable costs due to shutting down, firing up or changing the effective load of a power plant. Such ramp costs can be considered second order and would not have significantly affected final results. A new NEA study on the system effects of nuclear power that is currently under preparation considers in more detail the potential and cost of load following by nuclear plants.

from a suspension option but in an overall limited manner that sees the suspension option generating between 16% and 18% of additional per MWh profits, depending on the carbon pricing regime.[6]

Table 5.2: The value of the ability to suspend production

	Carbon trading (EU ETS 2005-10)			Carbon tax (equal to average EU ETS price)		
	Coal	Gas	Nuclear	Coal	Gas	Nuclear
Average profit with suspension option (EUR/MWh)	16.99	14.48	39.48	16.98	14.22	39.48
Average profit without suspension option (EUR/MWh)	16.98	12.27	39.48	16.98	12.27	39.48
Value of suspension option (EUR/MWh)	0.01	2.22	0.00	0.00	1.95	0.00
Value of suspension option (%)	0.00	18.00	0.00	0.00	16.00	0.00

The suspension option instead is worthless for the technologies with lower variable costs, coal and nuclear. These will have to bear passively any period of low electricity prices as long as they are higher than their variable costs, which during the period of analysis was the case for almost 100% of the time. However, one should note that when testing for higher CO_2 prices, as will be done in Chapter 6, coal becomes the marginal fuel far more often and thus increases the value of its suspension option. This comes, of course, again at the price of decreasing its load factor, which due to its higher fixed cost is a larger burden to carry than in the case of gas with its very low fixed costs.

On a methodological level, calculating the suspension option with empirical data is straightforward. It suffices to calculate the difference of returns with and without suspension option, which constitutes the value of the latter.

As one would expect on the basis of the results of the preceding section, the value of the suspension option for gas is higher under carbon trading than under a carbon tax due to the increased volatility of profits. Interestingly, the mode of allocation, free allocation or auctioning in a carbon trading system does not have any impact on the absolute value of the suspension option (although its percentage value in terms of profits will change). This is due to the fact that producers will suspend production when their *opportunity cost* of production is higher than the benefit of production, i.e., the electricity price. This means that gas-based power producers will include the price of carbon in their decision to maintain or to suspend production, regardless of whether they paid for permits or received them for free. While the latter is much more profitable, it has no incidence on production decisions, since the freely obtained carbon permits now hold value in the carbon market (see Chapter 4 for a discussion of the notion of opportunity cost).

---

6.   In this analysis, the suspension option was calculated on the basis of average daily prices. Given that gas turbines have ability for short-term load following on an hourly basis, the true value of the suspension option might be somewhat higher. However, one should not overlook the fact that the option to suspend production is only available to the share of production that has not been negotiated in forward contracts and is traded on spot markets. It is, of course, conceivable that producers would cover such longer term commitments with baseload technologies such as nuclear and shorter term commitments with more flexible technologies such as gas. This would further increase the value of the suspension option for gas but lower its average revenue due to lower load factors. In the end, a precise answer would not require a technology-against-technology comparison but a portfolio approach. The former have the advantage of offering transparent answers to relatively simple questions, while the latter offer more circumstantiated responses driven partly by *ad hoc* assumptions to these more complex questions.

Chapter 6
Investment analysis

The investment analysis together with the carbon tax analysis is the centrepiece of the NEA study on the competitiveness of nuclear energy under carbon pricing. Contrary to the profit analysis which compared *ex post* the historical profits made during the past five years by existing power plants, the investment analysis now compares the total costs and benefits over the lifetime of new plants that will begin operations in 2015. It thus adopts the point of view of a private investor who has to decide today whether to invest in a gas, coal or nuclear plant in order to produce electricity over the next few decades. The context in which investors in OECD countries will make that decision is likely to be similar to the one experienced during the past five years, i.e., characterised by liberalised electricity, fuel and carbon markets.[1]

This forward-looking *ex ante* analysis necessarily requires a number of assumptions that are presented below. Such modelling exercises, of course, always allow for more than one set of reasonable assumptions. Some assumptions are straightforward, and unlikely to become subject of debate among informed observers; others are less straightforward but not necessarily critical for the final result. A third category of assumptions allows for more than one reasonable choice in the face of an uncertain future, but this choice may well have a significant impact on the outcome of the modelling exercise. Capital costs, the level of electricity, carbon and gas prices or the profit margin of the marginal fuel are such critical parameters. This is why the present study includes a number of sensitivity analyses to provide framing and context for the baseline results. The chapter thus has the following structure: Section 6.1 will introduce the methodology employed and Section 6.2 will present the base case, high electricity price and low electricity price scenarios.

6.1 Methodology

The investment analysis works with a combination of historical fuel price data from European electricity, carbon, gas and coal markets between July 2005 and May 2010 as well as with cost data from the IEA/NEA study on the *Projected Costs of Generating Electricity: 2010 Edition*. The Projected Costs study thus provided the costs for investment and O&M. The most important assumption concerns the use of the *historical* price data for the modelisation of *future* electricity, fuel and carbon prices. In other words, for a new nuclear plant with a lifetime of 60 years and to be commissioned in 2015, it is assumed that it will face the same electricity prices that prevailed during the 2005-10 period in 12 five-year increments.

---

1.   For the time being, carbon markets only exist in the European OECD countries and electricity markets in the Asian OECD countries have yet to be liberalised, while Canada, Mexico and the United States present a patchwork of infra-national markets with different regulatory structures. Given the availability of data and the projects for new nuclear power plants in Finland, France, Italy, Poland, Switzerland and the United Kingdom, the European situation is of particular interest. Nevertheless, the analysis also holds important lessons for the countries of OECD Asia and OECD North America. No matter where, implicit or explicit carbon pricing is more than likely to become a reality for any power plant coming on-stream in 2015. Broad-based advances in transmission and information technology will also facilitate the monitoring and communication of electricity flows in transmission, distribution and consumption and thus further the liberalisation of power markets.

Basing the analysis of future profits only on data from the May 2005 to June 2010 in the European power market may appear arbitrary, especially considering the high fluctuations that have been observed during recent years in electricity carbon and commodity prices. However, one needs to consider first and foremost that the 2005-10 period encloses all of the available data of carbon emissions pricing. In addition, the past five years encompass a period of high economic growth followed by a deep financial and economic crisis and thus provides a good cross-section of different economic conditions. There are also currently few indications that the dynamics determining price formation in electricity and carbon markets will drastically change in the future.

Nevertheless, using the past five years as the basis for predicting the following 60 years remains a bold assumption and can only be justified in the light of the alternative of explicit modelling electricity prices for the next 60 years. While some such modelisation has been attempted (see Geman, 2005; Yang and Blyth, 2007; Pozzi, 2007), the results are by and large unconvincing and of little use for long-term empirical analyses such as this one. The reasons are the following:

- the relatively short period during which liberalised electricity have existed makes reliable calibration of the forecasting equations difficult;

- modelling electricity prices in this case would have required modelling not only spot but also forward prices in order to derive the true level of returns for production;

- electricity spot prices are only partially driven by cost or price fundamentals but by short-term changes in demand due to very uncertain parameters (temperature, large sporting events, TV programmes, etc.) as well as the instantaneous monopoly power of the marginal producer; these are difficult to capture even by technically sophisticated "jump diffusion" models with untested long-term performance;

- none of the available models takes into account the formation of electricity prices in the context of carbon pricing.

Clearly, electricity, carbon, gas and electricity prices over the next 30, 40 or 60 years will not be precisely the same as those over the past 5 years. Nevertheless, this assumption might be considerably closer to future reality than any alternative. Taking the only available empirical data on electricity prices in the context of carbon pricing as the basis for future projections offers in fact a number of advantages:

- first and foremost, its transparency; readers who are familiar with the main characteristics of the price data from the profit analysis can easily draw their own conclusions on the basis of their own price expectations;

- the use of historic data maintains the correlations between electricity prices and other variables (carbon and fuel prices), whose explicit modelisation would pose a number of questions concerning their correlation for which no unequivocal answers exist;

- the profit margin, i.e., the difference between the electricity price and the variable cost of the marginal fuel, is maintained; explicit modelisation of this profit margin would require assumptions about complex and transitory relationships between capacity utilisation and monopoly power that are bound to be very arbitrary;

- finally, a look at Figure 5.1 shows how the past five years include a large array of different price levels as well as correlations between different prices; periods of fast growth alternate with recession and relative stability; while the probability that the whole series of events will be replayed is exceedingly small, the probability that different elements of the series will be repeated is rather high; this is also the reason why the investment analysis and the carbon price analysis work with "high" and "low" scenarios for electricity and gas prices.

Working with historical data as the basis for projections of future prices thus seems by far the most useful and practical assumption. Historical price data were also used for the "high" and "low" electricity price scenarios as well as the "high" and "low" gas price scenarios in Chapter 7. In these cases, the 12 months with the highest or lowest electricity or gas prices were selected to constitute the whole time series. Integrating the selected months wholesale, that is keeping *all* historical prices during the selected periods together, allows again maintaining the historical correlations and short-term dynamics. An explicit modelisation would again have required employing a large number of assumptions that are difficult to justify.

Concerning the cost data stemming from the Projected Costs study, the study takes the *mean* values of the European entries for nuclear, coal and gas plants in order to determine the costs for investment, operation and maintenance and decommissioning, as well as the appropriate coefficients for the efficiency of conversion and carbon emissions. This yields the parameters in Table 6.1.

Table 6.1: Assumptions on cost and technology

	Nuclear	Coal	Gas
Technical assumptions			
Capacity	1 443 MW	723 MW	526 MW
Construction years	7	4	2
Lifetime	60	40	30
Electrical conversion efficiency	n.a.	0.44	0.55
Gross energy content of fuel unit	n.a.	6.98 MWh/tonne	1 MWh
CO_2 emissions per MWh	0	0.78 tCO_2/MWh	0.37 tCO_2/MWh
Cost assumptions			
Overnight costs[a]	3 291 EUR/kW	1 898 EUR/kW	851 EUR/kW
O&M	10.57 EUR/MWh	5.9 EUR/MWh	3.54 EUR/MWh
Fuel[b]	6.59 EUR/MWh	Daily	Daily
Decommissioning	494 EUR/kW	95 EUR/kW	43 EUR/kW

a. Overnight costs refer to the first-of-a-kind case, see text below for explanations.
b. Fuel costs for nuclear energy include cost for the back-end of the fuel cycle, i.e., spent fuel disposal.

Source: IEA/NEA (2010), mean values of submissions from European OECD countries.

In addition to the assumptions specific to each technology, the study contains of course a number of generic assumptions that are common to all three of them. These concern the discount rate, the rate of technical availability and the length of operation per year. For the discount rate, which is assumed to be equal to the cost of capital, a rate of 7% real is taken. The generic rate of technical availability that determines the load factor is 85%. It differs from the load factor due to the suspension option, which was considered to be always available. Whenever the suspension option is exercised, the load factor is reduced.[2] Annual operating time was assumed to be 8 760 hours per year.

---

2.   At the level of the mechanics of calculation, the NEA model does not work with the load factor *per se* but with a capacity reduced by the factor of technical availability. In addition, production then stops each time the suspension option is exercised. The suspension option in the model was exercised by gas-fired power plants, the only ones concerned in a significant manner, on 8.9% of all trading days for 85% of total capacity. This means that exercising the suspension option reduced the load factor by a further 7.5%. The effective load factor of gas-fired power plants in the model is thus 77.5%.

The most critical assumption concerns of course the discount rate. In order to ensure transparency and readability as well as to be able to concentrate on different aspects such as different price scenarios, the study works with one single discount rate. The issue of discount rate sensitivity was widely discussed in the Projected Costs study that used the two real discount rates of 5 and 10%. A discount rate of 7% seems a reasonable compromise rather close to the true cost of capital of large European utilities.[3]

First-of-a-kind case and industrial maturity case

An additional issue arises from the other component of investment costs, the overnight costs, or the costs of construction net of the interest payments due during the construction period. This indicates to some extent the efficiency of the plant vendor and crucially determines the competitiveness of the different technologies. It is quite obvious that for large, technically complex industrial installations such as power plants there exists a considerable difference in the overnight costs for the first plant ever being built, the so-called first-of-a-kind (FOAK) plant and the nth plant of a series of plants.

This issue plays a major role in the sample for nuclear power plants under consideration in this study. While coal and CCGT gas plants can be considered mature technologies with by and large quite predictable costs, seven out of the ten nuclear plants provided by European member countries in the Projected Costs study refer to advanced Generation III+ reactors.[4] Several reactors of this generation are currently being built but none of them has yet been connected to the grid. The cost estimates provided in the Projected Costs for commissioning in 2015 thus clearly refer to FOAK plants, i.e., plants for which no prior construction experience could be gained. This is why this study introduced a so-called "industrial maturity" case that assumes that generation III+ reactors, in Europe as elsewhere, will benefit from the economies of scale due to increased experience with increased installed capacity. Such economies of scale have been analysed in the form of "learning curves", which express the relationship between FOAK costs, installed capacity and construction costs in the following manner:

$$\text{Cost}_N = \text{Cost}_{FOAK} * \text{TICAP}(N)^\alpha, \alpha < 0$$

where the cost of the nth plant is equal to the cost of the FOAK plants multiplied by the total installed capacity (TICAP) to the power of the constant learning elasticity α, with α being negative.[5] For ease of exposition and comparison, learning rates are frequently expressed with respect to a doubling of capacity. 2^α is then referred to as the progress ratio (PR). The learning rate (LR) itself is then the complement of the progress ratio or

$$LR = 1 - PR = 1 - 2^\alpha.$$

If one applies a learning rate of 10% to the overnight costs of European power plants, i.e., a decrease of overnight construction costs of 10% with every doubling of production, this would imply a progress ratio of 90% and a learning elasticity, α, of -0.15. This in return implies that the construction of 14 power plants would bring down costs to two thirds of the original first-of-a-kind costs. Given that worldwide about a dozen of generation III+ reactors are currently under construction this

---

3.   Cambini and Rondi (2010) report a *nominal* WACC of 7% for a large sample of European utilities during 1997-2007. While ongoing liberalisation has probably increased the WACC, adjusting for inflation would still suggest a *real* rate of around 7%.

4.   This is different for the cost estimates for nuclear power submitted in the context of the Projected Costs study for OECD North America and OECD Asia, which are based on alternative technologies, several of them well known since decades. This largely explains the difference in the relative performance of nuclear power between the three OECD regions (see IEA/NEA, 2010, pp. 18-19).

5.   This exposition of learning curves follows Rogner and McDonald (2008), p. 87.

is far from unrealistic. A one-third reduction in overnight cost was thus retained in the industrial maturity case that is being presented alongside the first-of-a-kind case in the base case scenario and employed in the scenario analyses.

Box 6.1: How realistic are the FOAK case and the industrial maturity case after Fukushima?

The first-of-a-kind (FOAK) case and the industrial maturity case can be interpreted as the upper and the lower bounds of the future cost of the investment costs for nuclear energy, since the range encompasses designs currently under construction as well as being considered for further construction in the near future. The precise cost of future reactors will be difficult to determine for some time for two reasons. First, deployment of the new Generation III and III+ reactors will generate some economies of scale, but how much is difficult to say. Second, the partial fuel meltdown at three nuclear plants after the failure of the cooling systems in the wake of a major earthquake and a large tsunami at the Fukushima nuclear site in Japan will trigger a regulatory review of the safety features that will be requested for existing as well as new nuclear power plants. It is too soon to draw conclusions on the cost implications of the requirements emanating from the lessons learnt at Fukushima. While there will be some impact in terms of added costs, there is reason to think that it might be limited given that Generation III reactors already have a number of safety features such as multiple (up to four) independent cooling systems, including passive cooling, core catchers and outer containment domes (in addition to the interior reactor containment vessel) able to withstand high pressures. In other words, even after Fukushima, the first-of-a-kind case is likely to remain a valid upper bound for new European nuclear reactors.

Which measure for the profitability of investments?

A key question for the investment analysis concerns the methodology to be used to measure the profitability over the lifetime of each project. In order to assess the profitability of different technologies, this study adopts the perspective of a private investor who has to commit funds for one, and only one, power generation project for commissioning in 2015, the year for which the data in the Projected Costs study were provided. The investor will choose the technology that is likely to award his/her investment with the greatest return. Clearly, the LCOE methodology that was successfully employed in the Projected Costs study is no longer appropriate since the present study works with historical, i.e., exogenous, price data.[6]

A logical alternative for assessing the profitability of different technologies is cost-benefit accounting resulting in the assessment of the NPV over the lifetime of an investment. It is a robust and intuitively appealing measure that can easily handle exogenous price data. This study does provide NPV results but it should be clarified immediately that this is just done in first approximation and that NPV is not considered the appropriate measure of profitability in this study. This is due to the fact that absolute NPV results are largely dependent on project size and independent of the relative profitability of the initial investment. Even a marginally profitable nuclear plant is thus likely to have a higher NPV than a highly profitable gas plant.

---

6.   The LCOE methodology yields as its result the constant price of electricity at which a power plant would break even. By construction, the electricity price is thus both endogenous and stable. The LCOE methodology provides no information about the level of profitability obtained once observed electricity prices exceed the calculated break-even level and is thus unsuited to deal with volatile prices.

Of course, there exist situations where absolute NPV calculations are still appropriate. Imagine a situation, say on a peninsula, on which there is room for one single plant but the market could still absorb any amount of power. In this case, building a 1 600 MW nuclear plant is preferable to a 400 MW gas plant, both from a social planning as well as a private investor point of view. However, the philosophy of this study is that investors can choose any number of plants of the most profitable alternative in an open, unconstrained market. Thus NPV needs to be normalised in order to compare investments of different sizes.

Absolute NPVs can be normalised according to two parameters, plant size or investment costs. While normalising for plant size would have been an acceptable alternative, this study chose invest-ment costs as the denominator for relative NPVs, mainly because it is closest to the concern of private investors to maximise the value of their investment. The so-called profitability index (PI) usually indicates the ratio of discounted revenue (discounted cash flow or total present value) over investment costs (overnight costs plus interest during construction) as a measure of profitability. The PI, also known as the "benefit-cost ratio", is an established measure for the ranking of different investments and provides a particularly clear answer to the guiding question of the private inves-tor "which is the investment in which, Euro per Euro, I am obtaining the greatest return?" Much of the remaining chapter, in particular the scenario analyses, thus concentrates on comparing the competitiveness of nuclear, coal and gas plants on the basis of their profitability indices, which are usually defined as:

$$PI_{Standard} = TPV/INV,$$

where PI is the profitability index, TPV the total present value of the project including investment costs (equal to NPV plus investment costs) and INV are investment costs. If the PI is greater than one, the investor is making a positive return and the investment is worth undertaking. This study uses a slightly transformed version of the profitability index that uses the fact that net present value is the difference between total present value and investment:

$$NPV = TPV - INV.$$

It thus arrives at a formulation that maintains the link with the previous net present value cal-culations and emphasises the ranking between different projects:

$$PI_{Study} = NPV/INV = TPV/INV - 1 = PI_{Standard} - 1.$$

In this study an investment is thus profitable when the PI is positive. In order to present a com-plete picture of the comparative profitability of nuclear, coal and gas plants in European electricity and carbon markets, the study also provides results for the calculations of the modified internal rate of return (MIRR), a variant of the widely used internal rate of return (IRR). The IRR is again an endog-enous measure, which indicates the cost of capital at which a given sum of positive and negative cash flows would render the NPV equal to zero:

$$NPV = \sum^{N} \text{Net income}_n/(1+IRR)^n = 0,$$

where N is the lifetime of the investment in years (or any other appropriate unit) and n the par-ticular year in which the net income is generated. The IRR thus provides a hurdle rate with the help of which investors can decide whether their actual cost of capital is higher or lower than the IRR before undertaking the project. IRR calculations have the additional drawback of providing multiple solutions when large expenditures occur during the lifetime of projects, such as refurbishments or decommissioning.[7] The modified internal rate of return (MIRR) avoids multiple solutions and allows for different assumptions about reinvestment rates. It is defined as:

---

7.   A final shortcoming of IRR calculations is the necessary, but frequently unrealistic, assumption that the reinvestment rate for funds is equal to the cost of capital. This particular assumption, however, would not have posed any problem in the present context.

$$MIRR = [(\Sigma^N Income_n/(1+RR)^n)/(\Sigma^N\text{-}Cost_n/(1+WACC)^n)]^{1/N} - 1.$$

In short, the MIRR results from taking the root (appropriate to the lifetime of the project, N) of the ratio of the sum of the positive cash flows (discounted at the reinvestment rate, RR) and the sum of negative cash flows (discounted at the WACC). In doing so the MIRR loses some intuitive appeal but remains a fairly robust measure of a project's profitability. In the end the results that it provides and that are reported in this study are very similar to those provided by the profitability index on which this study is concentrating.

A crucial difference between the historic profitability analysis presented in the previous chapter and the investment analysis in this chapter is that power providers are assumed to have to pay for their carbon emissions. In other words, CO_2 permits are no longer attributed for free (as was the case during 2005-10) but have to be acquired through payment, most likely through a series of government sponsored auctions. The price of the carbon permit is, of course, nevertheless assumed to correspond to the historically observed price due to the principle of opportunity cost (see explanations in Chapter 2). Once electricity companies are in possession of the permits, their value, price and correlation with other variables no longer depend on the mode through which the companies acquired them in the first place.

Scenarios for sensitivity analysis

In order to be able to provide a more complete picture of the impact of carbon pricing on the competitiveness of nuclear energy, the study also presents a number of scenarios in addition to the base case scenario. Naturally, they focus on the key drivers of the comparative profitability of nuclear, coal and gas, which are electricity prices, investment costs, carbon prices and gas prices. The different scenarios are grouped in three sections:

1. *Electricity price scenarios:* in addition to the base case scenario, the study presents two scenarios with high and low electricity prices. The high (low) electricity price scenario bases the investment analysis on the 12 months of the 2005-10 period with the highest (lowest) average electricity prices. For information, the average electricity price during the 2005-10 period amounted to EUR 55 per MWh, while the average during the 12 months with the highest prices was EUR 70 per MWh and during the 12 months with the lowest prices EUR 46 per MWh. The section on electricity price scenarios also provides an analysis of the impact of electricity price expectations assigning different probabilities to each of the three electricity price scenarios.

2. *Overnight investment cost analysis:* this section shows the great importance the size of the overnight investment costs of nuclear power holds for its competitiveness. Due to the high fixed-cost-to-variable-cost ratio of nuclear power, its overnight costs have an over-proportional impact on competitiveness compared to the overnight costs of coal or gas. In addition, there is some reason to assume that there is considerably more room for "learning" (see discussion above) during the construction of new Generation III+ reactors than for coal and CCGT gas plants, which are largely mature technologies.

3. *Carbon and gas price scenarios:* these scenarios are presented in Chapter 7 which attempts to answer the question "what would happen to the competitiveness of nuclear power if carbon prices increased in a market environment similar to the 2005-10 period?" The analysis works with a sliding carbon tax and considers high and low gas price scenarios together with the base case scenario.

With the help of these different scenarios, the study covers most of the relevant perspectives from which the competitiveness of nuclear power under carbon pricing might be approached. Like in any modelisation of the future, a number of results remain inevitably driven by assumptions. Nevertheless, by presenting the assumptions made at each turn in a comprehensive and transparent manner, the results allow the identification or confirmation of a number of important findings that are presented individually in the following section for the results of the investment analysis, in Chapter 7 for the carbon tax analysis and comprehensively in the conclusions of Chapter 8.

6.2 The investment base case and electricity price scenarios

The investment base case provides the NPV, the MIRR and the PI for nuclear, coal- and gas-fired power plants to be commissioned in 2015. Given the important impact of the level of electricity prices, the NPV, MIRR and PI are calculated separately for each electricity price scenario. We recall that the base case scenario reflects the price dynamics during the 2005-10 period by repeating prices for the different variables in five-year increments over the lifetime of the plant. The high (low) price scenario instead relies on the 12 months with the highest (lowest) electricity prices, repeating the values in one-year increments over the lifetime of the plant. The NPV itself is calculated in the following manner:

$$\text{NPV} = -\text{INV} + (\textstyle\sum^N \text{Net income}_n/(1+r)^n) * Q - \text{DC}.$$

The NPV is thus composed of investment costs, the sum of annual discounted net income multiplied by annual production Q and decommissioning costs. The discount rate r in this study is equal to the cost of capital and, if not indicated otherwise, assumed to be 7% real. Annual production Q is assumed to be 7 446 MWh for every MW of installed capacity for all technologies and each year of a plant's lifetime (this number might be lower for gas due to the exercise of the suspension option). This corresponds to the 8 760 hours of the calendar year multiplied by a load factor of 0.85, which for simplicity is also assumed equal for the three technologies. Investment costs are calculated according to:

$$\text{INV} = \textstyle\sum^M (\text{share}_m/(1+r)^m) * \text{OC}.$$

Here M is the length of the construction period running from 0 to M with m being any particular year of construction, while share_m is the percentage share of overnight investment cost (OC) disbursed in year m and depends on the length of construction.[8] Overnight cost includes owner's cost, engineering, procurement and construction (EPC) costs as well as contingency costs net of IDC. Net income$_n$, the average net income in year n per MWh, is calculated analogously to the profit analysis in Chapter 5 as :

$$\text{Net income}_n = P_n - \text{O\&M}_n - \text{FC}_n - \text{CC}_n.$$

Here, P_n is the average electricity price in year n, O&M are operation and management costs, FC are fuel costs and CC carbon costs. The average electricity price, as well as the average fuel and carbon costs, are calculated on the basis of the real prices realised by electricity producers (either during the complete 2005-10 period, or its 12 months with the highest or lowest average prices), having the option to suspend production in case that the variable costs exceed the electricity price.

---

8.   In this formulation, the year M is considered to be 2015 for all technologies. For the construction times of 2, 4 and 7 years a linear distribution of construction expenditure was assumed.

On the basis of these calculations assuming a 7% real discount rate and the relatively higher capital costs of a FOAK nuclear plant, one obtains in Figure 6.1 the net present value of a nuclear, a coal and a gas plant under a base case scenario, a low price and a high price scenario. The implications of Figure 6.1 which shows the NPV to be generated over the lifetime of a power plant are quite obvious and foreshadow three key results of the analysis in this study that will be confirmed to different degrees under a variety of assumptions and from different perspectives over and over again. The first of these results is that a new coal plant is highly unlikely to be a competitive or even a profitable technology option under the price conditions prevailing during the 2005-10 period once it has to pay for its carbon emissions. We recall that the average carbon price during this period was slightly above EUR 14 and that the average coal price was EUR 63 per tonne.

Figure 6.1: Net present value in different electricity price scenarios
7% real discount rate, FOAK case and 2005-10 average carbon price

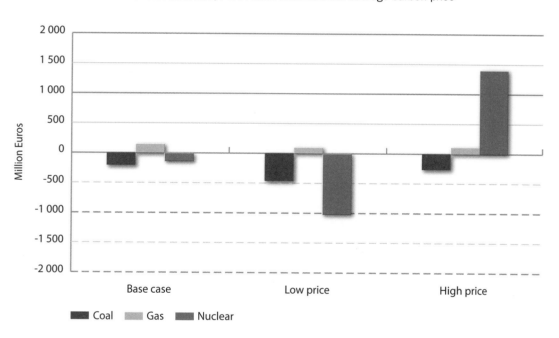

The second result is that the NPV of gas is relatively stable across the three different prices scenarios. This is primarily due to the fact that its sizeable variable costs are closely aligned with electricity prices, which limits downside as well as upside risk. The relatively small size of its fixed costs does not oblige it to generate very large profit margins in order to stay profitable. In addition, the suspension option allows gas to opt out of the market when prices are too low. High prices instead are not necessarily a source for significant additional profits as they frequently result precisely from high gas prices and consequently the high variable costs for gas-fired power plants.

The third result of the investment scenario is that the situation is precisely the opposite for nuclear power whose NPV depends almost exclusively on the level of electricity prices. Its high fixed costs and low and stable marginal costs mean that the profitability of nuclear rises and falls with electricity prices that single-handedly determine its profit margin, the difference between its per unit revenue and its variable costs. Given that the variable costs of nuclear power are virtually never above electricity prices and it thus has no opportunity to exercise the suspension option, nuclear power is bound to undergo electricity price changes in a largely passive fashion. Of course, the above results are based on absolute NPVs, which means that plant size matters. Other results based on measures that normalise for plant size and that are reported below, however, confirm these first findings.

The enormous importance of electricity prices and their expectations, given that the present investment analysis is formulated from the viewpoint of a private investor who has to make an investment decision in an uncertain environment, is also brought out in Figure 6.2. Here the NPVs of the three technologies are weighted as a function of the probabilities of the different price scenarios. Assuming a 33% probability for the base case scenario with average 2005-10 prices, the x-axis indicates different probabilities for the high price and the low price scenario. From left to right, the probability of the high price scenario thus increases from zero to 67%, while the probability of the low price scenario decreases at the same time from 67% to zero. While the NPV of a gas-fired plant is barely affected by this shift and the NPV of a coal-fired plant is only slightly affected, the NPV of a nuclear plant is very strongly affected and its competitiveness against gas depends very much on the expectations about electricity prices.

Figure 6.2: Expected NPV in function of the probability of a high electricity price scenario
7% real discount rate, FOAK case, 33% probability of base case scenario and 2005-10 average carbon price

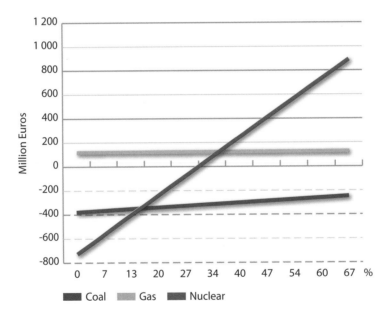

The situation changes fundamentally though if one moves from the first-of-a-kind case to the industrial maturity case. If overnight investment costs could indeed be reduced for a Generation III+ reactor in Europe by one third, then as shown in Figure 6.3 nuclear power would generate the highest absolute NPV in all three price scenarios. Of course, these findings will be put in perspective by normalising for plant size (see Figures 6.7-6.10), but the simple comparison of Figures 6.1 and 6.3 is quite instructive as to the importance of capital investment costs for the profitability of nuclear energy.

Figure 6.3: Net present value in different electricity price scenarios

7% real discount rate, industrial maturity case and 2005-10 average carbon price

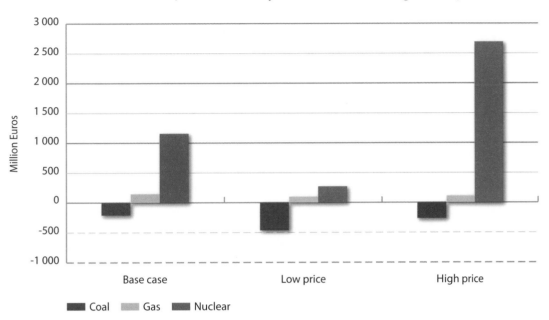

An even more powerful impact on the competitiveness of nuclear power is achieved by reducing the discount rate from 7% to 5% even in the first-of-a-kind case (see Figure 6.4). The discount rate reduction benefits, of course, all technologies. It is nevertheless particularly beneficial for nuclear energy. This is due to the fact that costs for nuclear energy are heavily front-loaded while benefits accrue over several decades up to 60 years, the end of the projected operating life of a modern nuclear plant. The lower the interest rate, the more valuable will be those future profits and the higher the overall NPV as long as prices hold up at reasonable levels.

In the set-up chosen for calculations in this study, which mimics the calculations a private investor might make who wants to start operating a plant in 2015, a reduction of the discount rate does not reduce overall investment cost (overnight costs plus interest during construction). Given that his/her decision will need to be made several years *before* commissioning in order to complete construction by 2015, a lower interest rate will actually *increase* the investment costs in his/her NPV calculation since the overnight costs will be discounted at 5% rather than at 7%.[9] This is indeed consistent with the point of view of a private investor deciding today which funds to commit in the future. It contrasts, however, with the results from the Projected Costs study, which took the day of commissioning rather than the day of the investment decisions as the reference point for comparing discounted lifetime costs, which meant that increased discount rates lead to significantly higher investment costs.

---

9.   This explains why the NPV in the low price scenario is slightly higher at a 7% discount rate for the industrial maturity case (Figure 6.3) than at a 5% discount rate for the first-of-a-kind case (Figure 6.4), although the NPVs are in the latter case much higher for the base case and the high price scenarios. In going from 7% to 5% in the low price scenario, the increase in operating profits due to a lower discount rate is in fact less strong than the increase in investment cost due to both the lower discount rate and the increase in overnight costs. The latter effect is swamped in the base case and the high price scenario by the very substantial increase in operating profits over the lifetime of the nuclear plant.

The difference between the choice of the date of commissioning or the date of the start of construction as the reference point is a virtual one. Going from one to the other means nothing else but sliding the value of all elements that enter into NPV calculations along a timeline, multiplying or dividing them by the appropriate discount rate. Moving from the date of commissioning to the date of the start of construction will thus lower investment costs in the case of a higher discount rates but it is important to understand that all future revenue will also be lowered in the same proportion and that final investment decisions will not be affected. Equally, the ranking between different technology options will be perfectly preserved.

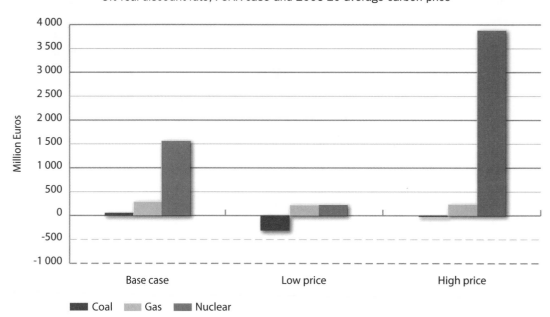

Figure 6.4: Net present value in different electricity price scenarios
5% real discount rate, FOAK case and 2005-10 average carbon price

What is important, of course, is that all technologies choose the same reference point for their calculations. In this case, the reference point is the start of construction of a nuclear plant, seven years before the date of commissioning, the moment the decision about the chosen technology has been made. Let there be no mistake: lower interest rates still unequivocally benefit nuclear in absolutely all cases. Even in the low price scenario the PI is considerably higher at a 5% interest rate (0.06) than at a 7% interest rate (-0.26) (see Figures 6.5 and 6.8). Since the decrease in interest rates lowers both investment costs and revenues, the difference between the two values decreases but the ratio of revenues over investment costs that defines the profitability index actually increases.

So far we presented results for absolute values of the NPV for projects based on different technologies irrespective of project size. The second part of the presentation of results for the base case concentrates on the *relative* profitability of different technologies normalised by project size. This is done by using the PI that provides the ratio of the NPV and the total discounted investment costs. The results must be understood in a manner that for say, a PI of 0.3, an investor will receive for every Euro invested EUR 1.30 in return over the lifetime of the project. Of course, he/she will receive much more in nominal terms over the lifetime of the project but this is the value of his/her investment at the very moment of investing, hence all future profits are properly discounted.

For the first-of-a-kind case and a real capital cost of 7%, the PI results in Figure 6.5 confirm the overall findings of the NPV analysis. With carbon pricing, coal is relatively uncompetitive in all pricing scenarios, gas is consistently competitive and the competitiveness of nuclear energy depends heavily on the level of electricity prices. In fact, if first-of-a-kind costs were a foregone conclusion and financing costs could not be reduced below 7%, European investors would choose nuclear power only if there was a significant probability of high electricity prices. In addition, even under the high electricity price scenario, nuclear and gas are almost at even level with even a very slight advantage for gas-based power generation.

Figure 6.5: Profitability index in different electricity price scenarios
7% real discount rate, FOAK case and 2005-10 average carbon price

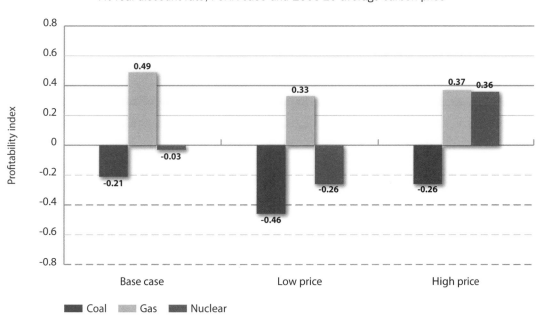

Very similar results are generated when employing an alternative manner to measure profitability, the modified internal return rate (MIRR) discussed above, the findings for which are reported in Figure 6.6.[10] In this case the results need to be interpreted in the following manner: when the MIRR is higher than the financing rate, the investment should go ahead, if it is lower, it should not be undertaken. Nuclear energy is thus a profitable proposition under the high electricity price scenario, although slightly less profitable than gas-fired generation. It is unprofitable in the low price scenario and just at the break-even point under the base case scenario. Investing in coal-fired power generation is never a profitable proposition and gas with its power to shape electricity prices is a profitable proposition under all three price scenarios under the assumption that future gas prices will not exceed the average of the gas prices observed during the past five years.

---

10. For comparison purposes one can consider that an MIRR of 0.08% at a cost of capital of 0.07% per year corresponds to a profitability index of 1.32 for a plant with a lifetime of 30 years and to a profitability index of 1.74 for a plant with a lifetime of 60 years.

Figure 6.6: MIRR in different electricity price scenarios

7% financing rate and reinvestment rate, FOAK case and 2005-10 average carbon price

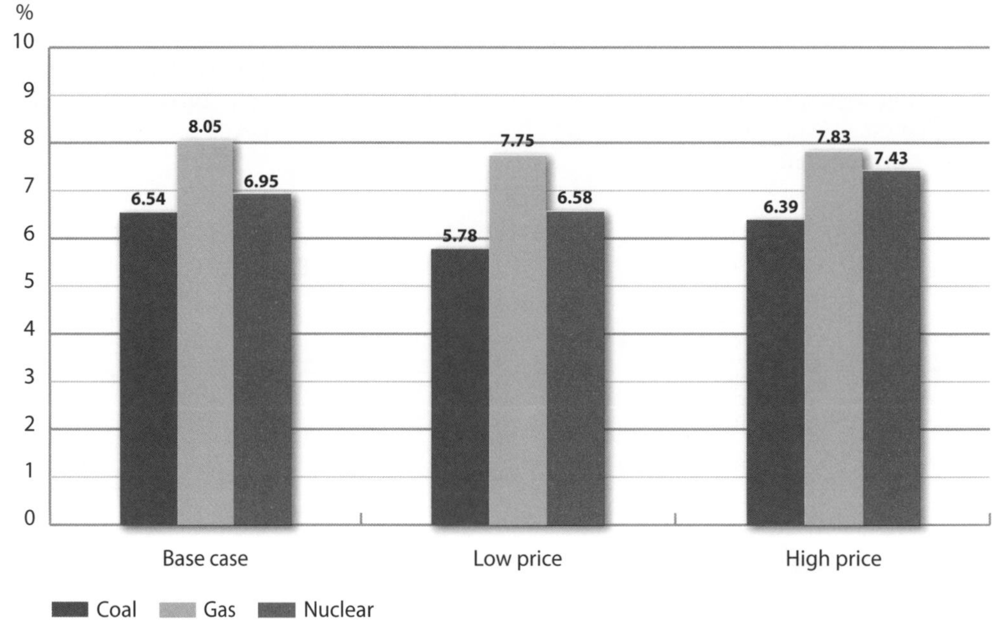

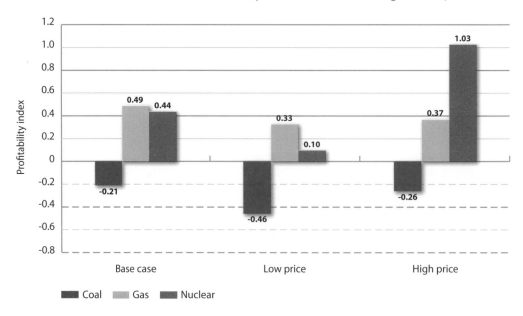

Figure 6.7: Profitability index in different electricity price scenarios

7% real discount rate, industrial maturity case and 2005-10 average carbon price

CARBON PRICING, POWER MARKETS AND THE COMPETITIVENESS OF NUCLEAR POWER, ISBN 978-92-64-11887-4, © OECD 2011

The situation changes again fundamentally when progressing from the FOAK case to the industrial maturity case with 67% of the original overnight investment cost of the median European plant from the Projected Costs study as pictured in Figure 6.7. While the situation for gas and coal does not change (first impressions due to a changed vertical scale notwithstanding), the profitability of an investment in nuclear energy improves markedly in all three price scenarios. Its profitability index is now a very respectable 0.44 (previously -0.03) in the base case scenario and even in the low electricity price scenario, which is unfavourable for nuclear energy, it manages to eke out a positive PI of 0.10 (previously -0.26).

Where the difference is most notable, however, is once more in the high electricity price scenario. There is simply no way around the insight that the profitability of nuclear energy as a high fixed cost and low variable cost technology rises and falls with electricity prices, an insight that should make nuclear energy a natural ally for efforts to improve energy end-use efficiency or increase carbon prices. If future electricity prices are at the level of the EUR 70 that correspond to the average electricity price of the 12 months with the highest prices during the 2005-10 period, investors would gain under the cost assumptions of the industrial maturity case more than double their outlay with a PI of 1.03 (previously 0.36), far above of what they would be able to gain with either coal- or gas-based generation.

Figure 6.8: Profitability index in different electricity price scenarios
5% real discount rate, FOAK case and 2005-10 average carbon price

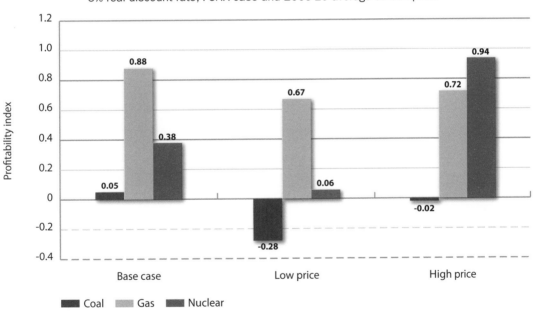

A similar but not identical effect would be achieved if the cost of capital could be reduced from 7% real to 5% real even if assuming the higher overnight costs of the first-of-a-kind case (see Figure 6.8). While the results for nuclear energy itself are absolutely comparable to the industrial maturity case at 7%, the impacts on competitiveness are not quite the same due to the fact that a decrease in the cost of capital would benefit all technologies and thus the profitability of both gas and coal would also increase. Of course, the increase would be of a lesser extent than in the case of nuclear energy due to the fact that nuclear as the most capital-intensive technology would benefit most from a reduction in financing costs.

Applying the relatively low real interest rate of 5% to the industrial maturity case, will of course further reduce investment costs and thus enhance the competitiveness of nuclear power (see Figure 6.9). Nuclear energy is now more than two-and-a-half times as profitable as gas-fired power generation in the high electricity price scenario, ahead of gas-fired generation in the base case and even coming close to competitiveness in the low gas price case. Clearly, this is a rather favourable set of circumstances for nuclear energy. However, it highlights once more that the destiny of nuclear energy depends only partly on the external circumstance of gas prices. To a substantial degree this destiny is in the hands of the nuclear industry itself. If overnight capital costs can be controlled and favourable financing terms arranged with long-term investors such as pension funds, nuclear energy remains clearly the overall most competitive option for power generation.

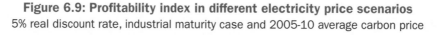

Figure 6.9: Profitability index in different electricity price scenarios
5% real discount rate, industrial maturity case and 2005-10 average carbon price

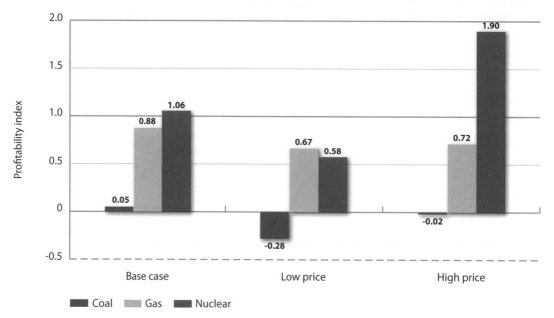

Drawing the conclusions of the results in Figure 6.9 allows once more to underline the importance of price expectations for investors faced with a choice between nuclear and gas, with coal being a largely uncompetitive solution under all price scenarios once it has to pay for its emissions. If base case expectations are held again at 33%, nuclear power becomes the most profitable option as soon as the likelihood of a high price scenario is 20% in the industrial maturity case with a 7% financing rate. Gas-fired power generation is again characterised by a relative independence from price expectations due to the already mentioned correlation of gas and electricity prices and low capital costs. The key conclusion of this first set of results is the importance of electricity prices and of overnight investment costs (see Figure 6.10).

CARBON PRICING, POWER MARKETS AND THE COMPETITIVENESS OF NUCLEAR POWER, ISBN 978-92-64-11887-4, © OECD 2011

Figure 6.10: Profitability index in function of the probability of a high electricity price scenario
7% real discount rate, industrial maturity case, 33% probability of base case and 2005-10 average carbon price

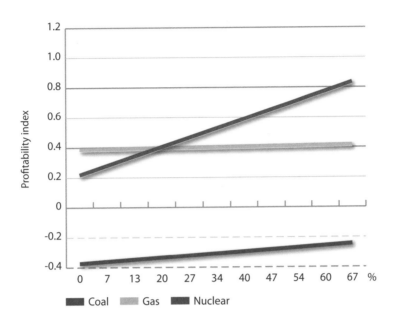

Figure 6.11: Profitability index (PI) in function of nuclear overnight costs
7% real discount rate, base case electricity price scenario and 2005-10 average carbon price

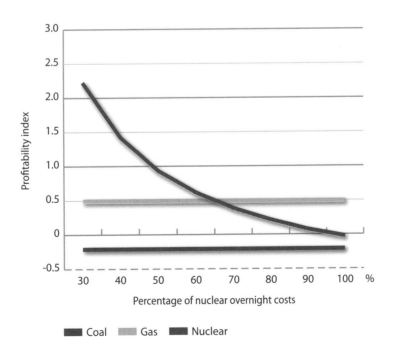

In conclusion, Figure 6.11 highlights once more the importance of overnight cost for the profitability and the competitiveness of nuclear power. With a 7% real cost of capital and in a base case scenario with an average electricity price of EUR 55 reflecting the cost and price conditions of the 2005-10 period, nuclear power becomes more profitable than gas only with a 30% reduction in overnight costs. Of course, reductions in the cost of capital, higher electricity, carbon or gas prices would all reduce the required efficiency gain. Since carbon and electricity prices frequently move up and down together there exists one, clearly defined electricity price scenario favourable for nuclear energy. Figure 6.10 is thus a stark reminder that the competitiveness of nuclear energy against gas is defined in the interplay between fixed costs, electricity and prices.

Chapter 7
Carbon tax analysis

In complement to the investment analysis presented in the previous chapter, the carbon tax analysis considers the central question of this study namely "what is the impact of carbon pricing on the competitiveness of nuclear energy?" While the previous chapter was built around the realities of carbon pricing in the EU ETS during the 2005-10 period with an average carbon price of around EUR 14 per tonne of CO_2, the present chapter will consider the impact of carbon prices evolving between zero and EUR 100 per tonne of CO_2. In other words, this chapter provides a glance into a future where carbon prices are likely to be substantially higher than today.

Chapter 7 thus highlights the importance of carbon pricing for the competitiveness of nuclear energy. While Chapter 6 already reported results for the impact of carbon pricing in an LCOE framework in the spirit of the study on Projected Costs (IEA/NEA 2010), this chapter looks at the impact of a carbon pricing on the basis of the empirical market data for the 2005-10 period and an extension of the NEA model already used for Chapters 5 (profit analysis) and 6 (investment analysis).

The results only partly confirm the intuition that higher carbon prices will substantially improve the competitiveness of nuclear energy in a liberalised electricity market. Of course, carbon pricing always has a positive impact on the profitability of nuclear energy due to the pass-through of higher carbon prices into higher electricity prices. The more surprising results concern the impact of higher carbon and electricity prices on the profitability of coal and gas. While the negative impact of carbon pricing on coal is unequivocal, the impact of very high carbon prices on gas-based power generation is, counter-intuitively, positive in the absence of carbon capture and storage (CCS). This is due to the fact that as carbon prices increase in a power market with liberalised prices coal becomes the fuel with the highest variable costs and thus the marginal generation technology which sets the electricity price. This, however, allows gas to earn additional "infra-marginal" rents that will be reflected in its profits. Since the rents of gas are modest in the absence of carbon pricing, its profitability grows in the absence of CCS very fast with the carbon price, faster even than that of nuclear, even though, of course, its own variable costs also increase.

A second noticeable fact that prevents the drawing of simplistic solutions is that the competition between nuclear energy and gas-fired power generation depends also heavily on the level of gas prices. In the low gas price case, for instance, the carbon tax required to equalise profitability is thus much higher than in the base case, while in the high gas price case, nuclear is more competitive even in the absence of carbon pricing.

7.1 The set-up of the carbon tax model

Such a modelling exercise necessarily requires again a number of important assumptions to be made, some of which may appear less justified than others, but all of which are necessary for the modelling to go ahead. The NEA model is again based on the empirical reality of energy markets during the 2005-10 period in order to preserve as much as possible the actual correlations between different variables. Data on key technical parameters remain unchanged from the previous chapter as summarised in Table 6.1. In the base case of the carbon tax analysis, prices for gas and coal will follow the day-by-day variations during the 2005-10 period, with the evolutions during these five years being scaled up for the duration of the lifetime of different plants. The basic set-up is thus identical to the model in the investment analysis in Chapter 6.

Regarding gas prices, also a high gas price and a low gas price will be presented. The high (low) gas price case will be based on the gas price series during the 12 months with the highest (lowest) gas prices, which are again then scaled up for the lifetime of the plant. The differences in gas prices between the three cases, base case, high gas price and low gas price case are quite remarkable testifying to the importance of gas price expectations in any investment decision in the electricity sector. While the average gas price in the base case is EUR 5.42 per MMBTU (this corresponds to EUR 3.64 per MWh of electricity produced), it is EUR 8.97 per MMBTU (EUR 55.63 per MWh) in the high price case and only EUR 2.87 per MMBTU (EUR 17.81 per MWh) in the low price case. It is obvious that such significant differences are bound to impact the competitive situation between gas-fired power generation and nuclear energy.

Concerning carbon prices, the varying carbon price of the EU ETS was substituted in the carbon tax model by a flat carbon tax rising at EUR 5 intervals to assess the evolution of competitiveness at different carbon price levels. Each point on the curves in Figures 7.1 to 7.13 in this chapter thus indicates a particular situation in which the corresponding carbon price is *constant over the lifetime* of the three plants. Modelling different carbon price levels as the result of progressively more severe carbon constraints in an emissions trading system would have yielded no additional insights and would have been considerably less transparent.

The greatest challenge in the modelling effort to assess the impact of different carbon tax levels is clearly the determination of the new electricity prices resulting from each distinct tax level. Higher carbon taxes will, of course, imply higher electricity prices. According to theory, and there is no reason to contradict theory on this point, electricity prices are a function of variable cost, which is composed of fuel costs, operation and maintenance costs and carbon costs. With higher carbon prices, the variable costs of gas and coal both increase gradually, the variable costs of coal increasing faster than those of gas. This logic in fact establishes a merit order between the different generation options and thus determines the price-setting fuel for each single day during the lifetime of the plants. It is easily verifiable that at low-carbon prices, gas and coal challenge each other for the spot as the marginal fuel. At higher carbon prices coal increasingly dominates the pricing process and installs itself almost permanently as the marginal fuel at carbon prices of around EUR 40 and higher.[1]

---

1. Gas will prevail as the marginal fuel even at carbon prices of EUR 100 per tonne of CO_2 only when taking the very high gas prices on a number of days during the first few months of 2006 when they almost reached EUR 100 per MWh at one point as the basis for calculations.

In modelling the link between carbon and electricity prices, the NEA model thus assumes a 100% carbon cost pass-through. Every increase in the carbon price, properly adjusted for emission factors, of course, will thus be reflected in the electricity price. As discussed earlier, this is a relatively common assumption that corresponds to the workings of a competitive market. Already during the 2005-10 period 100% cost pass-through was the default assumption due to the principle of opportunity cost (see Chapter 2). It is likely to be soundly confirmed once the allocation mechanism switches in 2013 from a free allocation of allowances to an auctioning mechanism with full payment for allowances.

The crucial question of mark-ups

To this point, the NEA model does not incorporate any assumptions that might be regarded as debatable. The most critical assumption by far concerns the profit margin or mark-up that is to be assumed as the difference between the variable cost calculated by the model and the electricity price. Real-world electricity prices, however, never precisely correspond to the variable costs of the marginal fuel. For the empirical 2005-10 data, the difference between average daily electricity prices and average daily variable cost of the marginal fuel differ between EUR 0.01 and EUR 32.81 per MWh with an average mark-up of EUR 13.82 per MWh. In other words, between nuclear power, coal-based and gas-based power CCGT plants even the marginal fuel made on average a EUR 13.82 profit per MWh for each MWh that it produced.[2]

The reasons for such mark-ups are multiple. To some extent they may indicate the costs of transporting coal or gas from the trading hub to the power plant costs, since fuel costs are calculated on the basis of prices at the physical wholesale markets which is net of the delivery costs to the plant.[3] Other errors will arise through the conversion of hourly values into daily values, given that on day-ahead electricity markets electricity is traded hourly. Another explanation is the existence of explicit or implicit market power that allows for prices above variable costs. The notion of "market power" is, of course, a very loaded term that needs to be carefully contextualised in electricity markets. In a market with non-storable goods, where supply and demand need to be matched instantaneously literally every second and suppliers communicate their production plans rather than their true production, "spontaneous" market power as opposed to consciously constructed market power can arise through any number of unforeseen events such as bottlenecks at interconnections or critical grid junctions, unexpected changes in the weather that lead to unexpected changes in supply (wind-power, hydropower) or demand (heating or cooling) or the non-anticipated impacts of one-off behaviour-changing events such as sporting events, election, TV programmes.

---

2.   All other, non-marginal fuels of course earn infra-marginal rents that correspond to the difference between the electricity price and their variable costs and that serve to finance their fixed costs.

3.   While it would be difficult to provide consistent Europe-wide figures for the transport of gas and coal, there exist data at national level that indicate that they are fairly low in comparison to the value of the fuel (less than 2% for gas and less than 3% for coal). The transport costs for gas would thus amount for a gas plant corresponding to the specifications in this study (see Table 6.1) to EUR 2.2 million per year or EUR 0.55 per MWh (see CRE, 2010). The transport costs for coal in Germany are estimated by Matthes at EUR 1.71 per MWh (Matthes, 2008).

Of course, the existence of such spontaneous market power does not preclude the existence and exercise of other, more traditional, forms of market power. In practice, however, it will be very difficult to tell the two apart. What is evident is that producers will plan their production in order to maximise any profit opportunities from upside demand risk and to minimise any exposure to downside risk. In practice, regulators and customers might not even be entirely opposed to such a practice as it provides some leeway for the cross-subsidisation of indispensable peaking capacity whose fixed costs would otherwise be very difficult to finance. In other words, some limited degree of monopoly power may contribute to the security of electricity supply. Finally, while this study on the profitability of different technologies for baseload power generation is based also on electricity prices for baseload, it is not excluded that at certain instances technologies other than those treated in this study have intervened in baseload power production.[4]

Box 7.1: What is the mark-up of electricity prices over the variable costs of the marginal technology?

The previous paragraphs provide a mixed picture. While there is no doubt about the existence of mark-ups over variable cost for the marginal baseload producer, its precise analytical determination is practically impossible on the basis of currently available data. It would be difficult in any case given the fact that such mark-ups are partly really due to "spontaneous" events. The data nevertheless show that mark-ups clearly reduce very quickly as carbon prices increase. For carbon prices below EUR 10 per tonne of CO_2, the average mark-up during the 2005-10 period was a massive EUR 19.92 per MWh, for carbon prices between EUR 10 and EUR 20 per tonne of CO_2, the average mark-up was EUR 13.32 per MWh and for carbon prices above EUR 20 per tonne of CO_2, the average mark-up was only EUR 9.76 per MWh. Clearly, mark-ups decline with carbon prices, which is fully consistent with microeconomic theory. As prices rise, consumer responses as expressed in the demand curve become more elastic vis-à-vis higher prices and profit-maximising utilities will reduce their margins in order to avoid excessive reductions in quantities. In principle this would indicate that there exists at least some degree of a conscious exercise of market power. However, answering this question in a more definitive manner would require much finer econometric study.

Of course, the figures above provide only very little information on mark-ups for carbon prices above EUR 30 per tonne of CO_2. The highest observed carbon price during 2005-10 was EUR 30.45 per tonne of CO_2, whereas the NEA model goes out to compare profitabilities for up to EUR 100 per tonne of CO_2. A linear regression analysis indicates that every increase in the price of carbon by EUR 1 reduces the mark-up by EUR 0.45. With an intercept of 20.30, this yields *negative* mark-ups for carbon prices higher than EUR 50 per tonne of CO_2, which is not a very likely proposition either. In the absence of a fully satisfying analytical solution, the NEA study chooses a simple default value of EUR 10 per MWh for all values of the carbon price between zero and EUR 100 per tonne of CO_2, fully aware of the preliminary nature of this choice.

The question of mark-ups, however, is crucial for the determination of the relative profitability of nuclear energy as compared to coal- and, in particular, gas-fired power plants. In fact the results of the analysis below show that other than the price of gas, the profitability of gas-fired power generation is almost entirely determined by the mark-up over variable costs that determines the electricity price. In order to understand this disproportionate impact of mark-ups over variable costs on the profitability of gas-fired power generation, one needs to recall that gas, since it is frequently the

---

4.   Rather than the sudden use of an open-cycle gas plant with very high variable costs, this should be thought of as the use of a particularly inefficient coal-fired power plant that has not been shut down after peak-load service due to ramp costs. Of course such a plant would have a different profitability than the one under analysis in this study but its use would still have impacts on the results for the latter.

marginal fuel, earns relatively little infra-marginal rents. The difference between its revenue and its variable cost, its profitability, is thus determined to a very large extent by the mark-up itself. It is able to survive quite nicely in this situation only due to its favourably low fixed costs. A change in the mark-up however will, positively or negatively, massively impact its overall profits (see Figures 7.1 and 7.2).

Nuclear energy is in precisely the opposite situation. Due to its low variable costs, it tends to earn very handsome infra-marginal rents even with relatively low mark-ups over the variable costs of the marginal technology. Even with mark-ups being wholly absent it would gain with every MWh produced, since it is almost never the marginal fuel. On the other hand, it very much needs those infra-marginal rents to finance its very large fixed costs. Its profitability will thus heavily depend on its fixed costs but to a much smaller degree on a variation in the mark-up. As mentioned in Box 7.1, this study finally chose a conventional mark-up of EUR 10 per MWh, which seems a reasonable compromise given the uncertainties surrounding the issue.

The particular cost structure of gas-fired power generation contributes also to the surprising fact that the profitability of gas-fired power generation *increases* with higher carbon prices, although gas does emit a non-negligible amount of greenhouse gas emissions, 0.37 tCO_2 per MWh of electricity compared to the 0.78 tCO_2 per MWh for coal in this study. While rising carbon costs increase the variable costs of gas, the variable costs of coal will rise much faster with the effect that at higher carbon prices coal is usually the marginal fuel. As shown in Section 7.2, this allows gas to earn additional infra-marginal rents especially at high carbon prices. The profitability of gas thus steadily rises with carbon prices, while the profitability of coal unequivocally decreases with carbon prices. Since the profitability of gas rises *faster* than the profitability of nuclear at high and very high carbon prices, the relative competitive position of nuclear does not necessarily improve at these high levels of carbon prices although its absolute profitability continues to increase.

It should be mentioned, however, that this only holds for markets with liberalised electricity prices in the absence of carbon capture and storage (CCS). CCS for coal plants, which for data reasons is only carbon capture (CC) in this study, changes the picture radically not only for coal itself but also for gas and to a lesser extent for nuclear. Once coal plants are equipped with CC, which reduces the carbon emission factor to 0.1 tCO_2 per MWh, gas becomes the marginal fuel at almost any carbon price. This reduces not only the profits of gas-fired power generation but also reduces electricity prices. As shown below, this means for nuclear energy that its absolute profitability declines (due to lower electricity prices) but its competitive position *vis-à-vis* gas improves. Despite a substantial improvement, coal cannot impose itself under the assumptions of the study even with carbon capture.[5]

Table 7.1: Discounted investment costs for different technologies
EUR per kW and 7% real interest rate

Nuclear	First-of-a-kind (FOAK)	3 913
	Industrial maturity (IM)	2 622
Coal		1 014
Gas		308

---

5.   This concerns primarily the assumptions for coal prices, which for month-ahead delivery in the Amsterdam-Rotterdam-Antwerp (ARA) market averaged EUR 63 (USD 93) per tonne of steam coal during the 2005-10 period. This is below long-term historical prices, but remains significantly below the price of EUR 88 (USD 129) per tonne in April 2011.

Results will be reported for the evolution of the profitability indices (PI) for nuclear, coal and gas. As explained in the previous chapter, the PI used in this study is defined as the ratio of net present value (NPV) over discounted investment costs (see Table 7.1). As mentioned above, since the NPV is the difference between discounted benefits and discounted costs (which include investment costs) any positive value of the PI indicates that the investment is profitable on its own. Needless to say, when one investment has a higher PI than another this means that the former is more profitable than another. This chapter does not report NPVs on their own since the relatively greater size of a nuclear plant always guarantees the top spot for nuclear energy in this metric. As indicated in the previous chapter, only the PI answers the question that interests private investors "which is the project in which I get the highest return on my original investment?"

7.2 Results for the standard carbon tax model

This section reports the results in terms of profitability indices of the standard carbon tax model for carbon taxes running from zero to EUR 100 per tonne of CO_2. In the standard model, nuclear power plants and gas-fired plants compete against coal-fired plants without any provisions for carbon capture and storage. The first three figures all reflect the base case, i.e., gas prices are assumed to correspond to the actual evolution of gas prices during the 2005-10 period.

Comparing Figures 7.1 and 7.2 shows the importance of the mark-up for the relative profitability of the three technologies, especially for those technologies like gas that do not earn any infra-marginal rents. Figure 7.1 thus shows a conceptual benchmark case for strict marginal cost pricing, i.e., a situation in which the marginal fuel, usually coal or gas (nuclear is the marginal fuel for all of 6 days during the five-year period between 2005 and 2010), does not earn any money above its variable costs for the electricity it produces. In this set-up, nuclear has a higher profitability than gas up to carbon prices of EUR 50 per tonne of CO_2 even under the relatively unfavourable assumption of FOAK capital costs. In the absence of carbon capture, coal is not competitive, and even with carbon capture it will rarely be the preferred option. Interestingly, however, coal is relatively more profitable than gas (but not more profitable than nuclear) in the absence of carbon prices and at very low-carbon prices. This is due to the fact that gas is more often the marginal fuel at low-carbon prices and is thus most affected by the absence of a significant mark-up.

Figure 7.2 shows the same configuration, i.e., also with first-of-a-kind investment costs for nuclear energy, but with a uniform EUR 10 profit margin over variable costs for the marginal fuel. It is immediately visible that this raises the profitability of *all three* technologies, including nuclear energy. This last effect is due to the rise of electricity prices that comes with a higher mark-up. However, while the profitability of nuclear increases only slightly, and that of coal only modestly, the profitability of gas increases substantially, in particular at low-carbon prices, where it was previously penalised by the absence of any mark-up. In fact, this effect is so strong that gas will most likely be the preferred technology over the whole range as long as the cost of capital, nuclear overnight cost or mark-ups remain at elevated levels.

As explained in Box 7.1, there is no firm theoretical basis for defining the mark-up in electricity markets, especially at carbon prices higher than the currently observed EUR 14, simply because the demand response to higher carbon and electricity prices is unknown. It is, however, quite certain that the level of the mark-up will have a very strong impact on the competitiveness between different power generation technologies.

Figure 7.1: Evolution of profitability indices in the base case scenario
Strict marginal cost pricing, 7% real discount rate and FOAK case

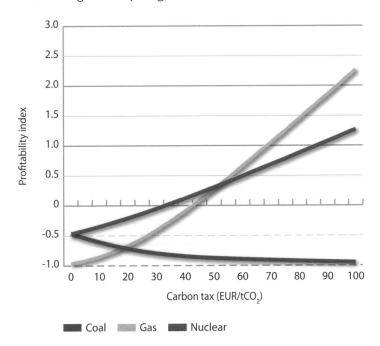

Figure 7.2: Evolution of profitability indices in the base case scenario
Constant profit margin of EUR 10, 7% real discount rate and FOAK case

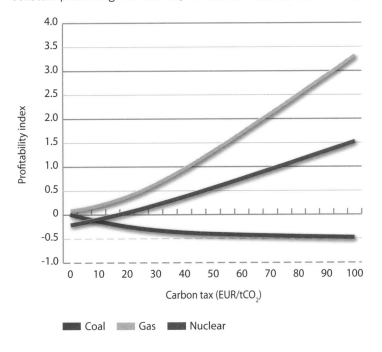

Figure 7.3a: Evolution of profitability indices in the base case scenario
Constant profit margin of EUR 10, 7% real discount rate and industrial maturity case

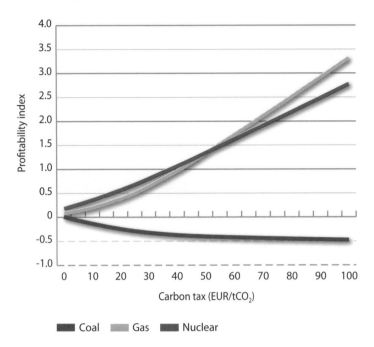

Figure 7.3b: Evolution of profitability indices in the base case scenario
Constant profit margin of EUR 5, 7% real discount rate and industrial maturity case

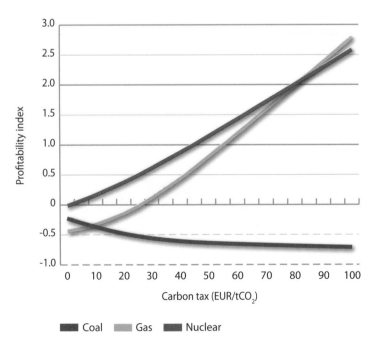

CARBON PRICING, POWER MARKETS AND THE COMPETITIVENESS OF NUCLEAR POWER, ISBN 978-92-64-11887-4, © OECD 2011

Once one moves to the industrial maturity case, where nuclear energy is able to reap a certain amount of economies of scale, the competitive picture changes again. While the profitability of gas and coal remains unchanged – since no marginal values are affected by the reduction in nuclear construction costs – the profitability of nuclear energy improves markedly. For carbon prices between zero and EUR 50 per tonne of CO_2, nuclear is now the preferred option. Given that Figure 7.3a reunites a number of realistic assumptions which make it a plausible reference case, this has potential policy implications. While it would be premature to insist on specific quantitative values (the uncertainty surrounding consumer behaviour at higher electricity prices and thus the uncertainty about mark-ups would not allow this), there is an important qualitative message contained in Figure 7.3a: there exists in fact a "window of opportunity" with respect to carbon prices, in which their contribution to the competitiveness of nuclear is highest.

The intuition that high and very high carbon prices will push investors automatically towards nuclear energy may have to be qualified. In fact, it would only hold if CCS plays an increasing role, if electricity market pricing will strictly follow marginal costs or if gas encounters other problems (such as security of supply issues). However, as has been pointed out before, in a pure market context with liberalised electricity markets, no supply constraints and mark-ups in line with historical precedent, gas will improve its competitiveness with high carbon prices as unconstrained coal sets high electricity prices as the marginal fuel. One should recall, however, that this assumes a "static" view of the state of technology. As pointed out above, carbon prices above EUR 50 per tonne of CO_2 will set in motion a number of factors such as coal without CCS leaving the market that would quickly put a cap on the profitability of gas at high carbon prices (see CCS analysis below).

However, even with gas prices remaining at historical levels carbon pricing will consistently ensure the competitiveness of nuclear energy over the whole range of politically sustainable levels of carbon prices as soon as mark-ups over variable costs are reduced (Figure 7.3b). The implications in terms of competition policy are straightforward. The competitiveness of nuclear energy against gas and coal would benefit from an opening of power markets, more competition in the provision of baseload power generation and reduced profit margins. As long as marginal producers with high variable but low fixed costs have the benefit of substantial profit margins, the competitiveness of nuclear energy will remain constrained. Removing those surplus profits, of which at least a share is due to spontaneous or voluntary monopoly power, will quickly re-establish the competitiveness of nuclear.

When progressing from the base case scenario to the low gas price and high gas price scenarios in Figures 7.4, 7.5 and 7.6, it becomes obvious how important the gas price is for both the absolute and the relative profitability of nuclear energy. The impact on absolute profitability of course passes through the electricity price which follows the gas price. The base case in Figures 7.1, 7.2, 7.3a and 7.3b corresponded to an average gas price of EUR 5.42 per MMBTU or EUR 33.64 per MWh of electricity during the 2005-10 period. Taking a gas price of just EUR 2.87 per MMBTU (EUR 17.81 per MWh) corresponding to the 12 lowest months during that period shows that nuclear energy is not competitive against gas at any carbon price. This effect will even supersede any cost reduction due to the lower overnight costs corresponding to the industrial maturity case (see Figure 7.4).

The opposite is the case when working with a high gas price of EUR 8.97 per MMBTU (EUR 55.63 per MWh) corresponding to the 12 highest months during the 2005-10 period. In this case, nuclear energy is the most profitable technology up to carbon prices of EUR 70 per tonne of CO_2 even when assuming the high overnight costs corresponding to the FOAK case (see Figure 7.5).

Figure 7.4: Evolution of profitability indices in the low gas price scenario
Constant profit margin of EUR 10, 7% real discount rate, FOAK and industrial maturity cases

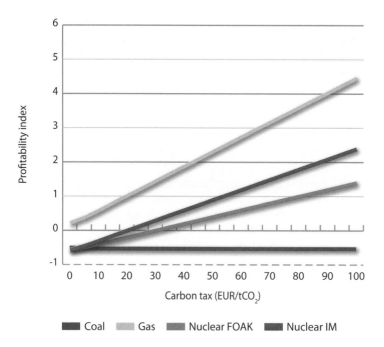

Figure 7.5: Evolution of profitability indices in the high gas price scenario
Constant profit margin of EUR 10, 7% real discount rate and FOAK case

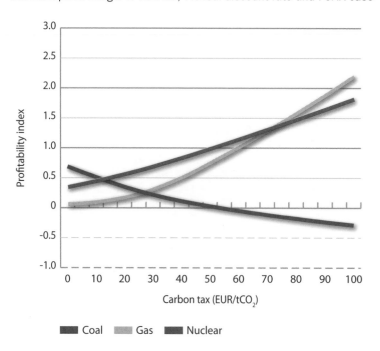

CARBON PRICING, POWER MARKETS AND THE COMPETITIVENESS OF NUCLEAR POWER, ISBN 978-92-64-11887-4, © OECD 2011

Figure 7.6: Evolution of profitability indices in the high gas price scenario
Constant profit margin of EUR 10, 7% real discount rate and industrial maturity case

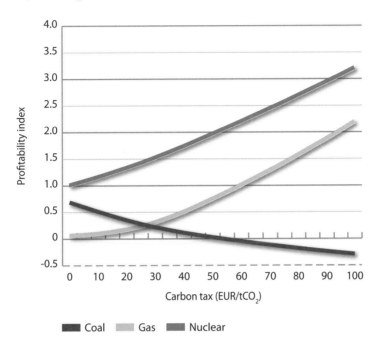

Its profitability and competitiveness, of course, only increase when moving to the industrial maturity case, where nuclear energy is always the most profitable technology even in the absence of all carbon pricing as shown in Figure 7.6. The flippant remark by the high executive of a European utility with substantial nuclear production that they had the same interests as Russian gas exporters, namely "high gas prices" bears more than a kernel of truth. The impact of higher gas prices on electricity prices not only lowers the relative profitability of gas but also directly increases the profitability of nuclear and coal through the impact on electricity prices. It is worth highlighting that the high gas price case is the only configuration in which coal is more profitable than gas, albeit less than nuclear, as long as carbon prices do not exceed EUR 25 per tonne of CO_2.

The results in this section which assumes that coal-fired power plants without carbon capture equipment determine electricity prices at medium to high carbon prices can easily be summarised as follows. The profitability and competitiveness of nuclear energy depends in roughly equal parts on:

1. Reducing overnight costs to progress from a first-of-a-kind scenario to an industrial maturity scenario.

2. A floor under gas prices; the competitiveness of nuclear against gas declines rapidly with falling gas prices, which almost single-handedly determine the profitability of gas.

3. Significant but not overly high carbon prices, since at very high carbon prices the profitability of gas improves disproportionately. This no longer holds with the introduction of pervasive carbon capture for coal-fired power plants. The next section will show that in this case, both high and very high carbon prices will improve the relative competitiveness of nuclear.

7.3 Results for the CCS carbon tax model

The previous section showed that due to its high CO_2 emissions per unit of output coal without carbon capture dominates electricity price setting at high and very high carbon prices.[6] It is thus logical to ask what would happen if pervasive equipment with CC would drastically reduce the CO_2 emissions from coal-fired power plants, making gas the marginal fuel most of the time, in particular at higher carbon prices. For this to happen it would, however, not suffice to equip just a sizeable portion, say 50%, of coal-fired power plants with carbon capture equipment. In such a configuration, coal-fired power plants without carbon capture would still be the marginal plants and set electricity prices. Thus coal plants with CC would just earn additional rents, as would nuclear and, in particular, gas. Plants with carbon capture equipment would truly need to be "pervasive" to the extent that coal-fired power plants without CC would only be drawn upon during peak times but would no longer intervene in the setting of prices for baseload power. Table 7.2 shows the assumptions derived from the Projected Costs study corresponding to the mean values of plants with carbon capture projected to be commissioned in Europe in 2015.

Table 7.2: Assumptions on cost and technology for coal-fired power technologies

	Coal	Coal with CC
Technical assumptions		
Capacity	723 MW	613 MW
Construction years	4	4
Lifetime	40	40
Electrical conversion efficiency	0.44	0.38
Gross energy content of fuel unit	6.98 MWh/tonne	6.98 MWh/tonne
CO_2 emissions per MWh	0.78 tCO_2/MWh	0.10 tCO_2/MWh
Cost assumptions		
Overnight costs	1 898 EUR/KW	3 114 EUR/KW
O&M	5.90 EUR/MWh	10.10 EUR/MWh
Fuel	Daily	Daily
Decommissioning	95 EUR/KW	156 EUR/KW

Source: IEA/NEA, 2010.

With pervasive CC technology the impact on prices and profits would be quite dramatic, improving the absolute and relative profitability of coal-fired power plants but *reducing* the absolute profitability of both nuclear energy and gas due to overall lower electricity prices. The most dramatic impact, however, is on the relative competitiveness of nuclear and gas, the latter's profitability declining massively at higher carbon prices once coal is no longer the marginal fuel due to carbon capture. Figure 7.7 shows how electricity prices are on average considerably lower once coal-fired power generators capture their CO_2 emissions, especially at higher carbon prices. The fact that electricity prices are slightly higher *with* CC equipment at very low-carbon prices is due to the fact that in this case the variable costs net of carbon costs, i.e., fuel and O&M costs, of coal-fired power generation with CC are somewhat higher due to reduced conversion efficiencies and higher maintenance costs.

---

6. While the precise point at which coal will dominate price setting depends on a number of specific assumptions, most notably the price of gas, one may think of "high and very high" carbon prices as prices above EUR 50 per tonne of CO_2 and more.

Figure 7.7: Average electricity prices in function of carbon tax and CCS

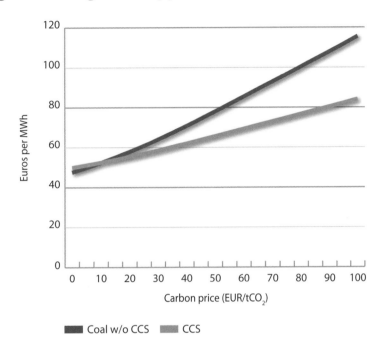

Figure 7.8: Evolution of profitability indices in the CCS base case scenario
Constant profit margin of EUR 10, 7% real discount rate, FOAK case and coal with carbon capture

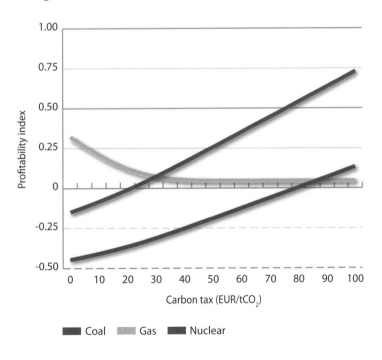

The impact of the switch towards carbon capture on profitability is brought out in Figure 7.8. In the base case with average gas prices, nuclear energy bypasses gas-fired power generation at carbon prices of around EUR 30 per tonne of CO_2 even with the high overnight costs of the first-of-a-kind case and increases its advantage from thereon. Coal-fired power generation with CC becomes competitive with gas at carbon prices of EUR 85 per tonne of CO_2. However Figure 7.8 should be compared to Figure 7.2 above which shows the corresponding case without carbon capture. While the relative competitiveness between nuclear and gas is reversed, the *absolute* profitability of nuclear has declined in Figure 7.8 except for very low-carbon prices. In the CCS scenario, a private investor would prefer nuclear energy to gas, but once the investment has been made he/she would prefer the absence of CCS.

Figure 7.9, which should be compared to Figure 7.3a, allows drawing similar conclusions. If investors prefer nuclear with overnight costs corresponding to the first-of-a-kind case once carbon prices exceed EUR 30 per tonne of CO_2, they will do so over practically the whole range once overnight costs correspond to the industrial maturity case. Since fixed costs by definition do not intervene in the formation of prices, the profitability of either coal- or gas-fired generation is not affected by the reduction of nuclear overnight costs.

Figure 7.10 shows that the switch to carbon capture also dramatically curtails the advantage of gas-fired power generation in a scenario of lower gas prices. In Figure 7.4, which showed the same configuration without carbon capture, gas was far ahead, its profitability steadily rising with carbon prices due to higher and higher electricity prices that were set by coal. This time, gas itself is increasingly setting the electricity price, being the only marginal fuel for carbon prices of EUR 35 per tonne of CO_2 and above. In this case, gas will only be gaining the mark-up of EUR 10 per MWh. It is easy to see that it would not be profitable at all in the absence of any mark-up or even somewhat smaller mark-ups. At low gas prices, nuclear energy becomes more profitable than gas at carbon prices of EUR 70 and above. It should be noted that overall profitability is rather low for all three technologies due to the low electricity prices set by gas, in itself not very expensive in this case, as the marginal fuel.

The low profitability of nuclear energy is, of course, mitigated in the industrial maturity case, where nuclear regains competitiveness against gas, which continues to benefit from low gas prices, already at carbon prices of around EUR 35 per tonne of CO_2 (see Figure 7.11). Again the profitability of either coal or gas is not affected as electricity prices would not change.

Finally, in the high gas price case combined with carbon capture for coal as shown in Figure 7.12 nuclear is by far the most profitable technology at any carbon price and even in the absence of carbon pricing and even when assuming the high overnight costs of the first-of-a-kind case. This is due to the fact that electricity prices are much higher at any single level of a carbon tax. One may recall from Chapter 6 the extent to which the profitability of nuclear depended precisely on the level of electricity prices due to its high fixed costs. Coal with carbon capture benefits from the same effect and will eventually become profitable but its overall cost structure (high fixed costs due to carbon capture, significant fuel costs as well as some remaining exposure to carbon prices) is not favourable enough in order to make it a truly competitive option.

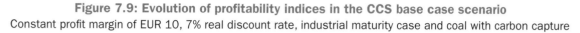

Figure 7.9: Evolution of profitability indices in the CCS base case scenario

Constant profit margin of EUR 10, 7% real discount rate, industrial maturity case and coal with carbon capture

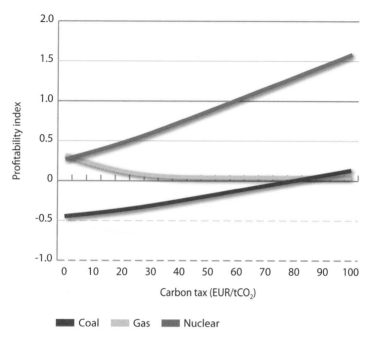

Figure 7.10: Evolution of profitability indices in the CCS low gas price scenario

Constant profit margin of EUR 10, 7% real discount rate, FOAK case and coal with carbon capture

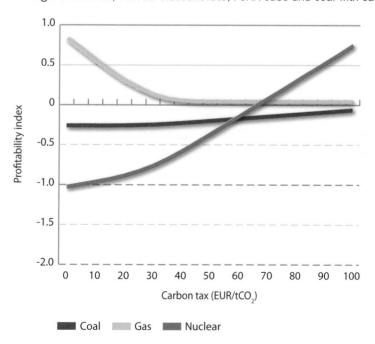

Figure 7.11: Evolution of profitability indices in the CCS low gas price scenario

Constant profit margin of EUR 10, 7% real discount rate, industrial maturity case and coal with carbon capture

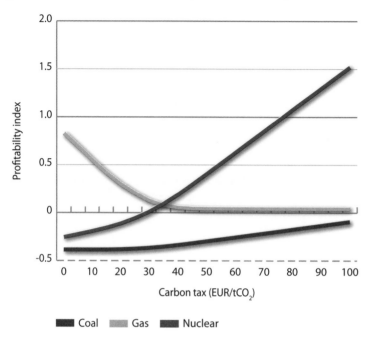

Figure 7.12: Evolution of profitability indices in the CCS high gas price scenario

Constant profit margin of EUR 10, 7% real discount rate, FOAK case and coal with carbon capture

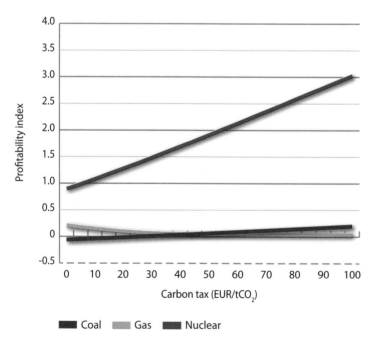

CARBON PRICING, POWER MARKETS AND THE COMPETITIVENESS OF NUCLEAR POWER, ISBN 978-92-64-11887-4, © OECD 2011

The same configuration, average profit margin of EUR 10 per MWh, 100% carbon capture for coal and a high gas price scenario, for the industrial maturity case yields evidently the same picture with only the profitability of nuclear energy further enhanced over the whole range of carbon prices (not shown). Needless to say, while this might seem a very comforting concluding picture for the competitiveness of nuclear energy, it assumes the coincidence of a number of rather favourable assumptions, such as substantial capital cost decreases, high gas prices and no more coal-fired power plants without carbon capture.

One thus needs to add to the three factors determining the profitability and competitiveness of nuclear energy mentioned in the conclusion of the previous section – reducing overnight costs, significant carbon prices and a floor under gas prices – a fourth one, the systematic installation of carbon capture equipment. Far from being an unwanted competitor, pervasive carbon capture has the potential of being a major element of ensuring the relative competitiveness of nuclear energy by significantly limiting the profitability of gas-fired power generation. While the downward pressure that carbon capture exerts on electricity prices limits also the absolute competitiveness of nuclear energy to some extent, its absence requires a number of rather favourable conditions in order to sustain direct competition with gas in a liberalised electricity market.

Clearly, the above analysis is based on the technological parameters of today and carbon prices of EUR 50 and more would generate a number of "induced" technological changes whose direction and magnitude is difficult to predict. Nevertheless, to the extent that they rely on empirical data for prices and technical assumptions the above findings constitute a robust first orientation for the impact of carbon pricing on the competitiveness of nuclear power for the generation of baseload electricity in liberalised power markets.

Chapter 8
Conclusions

This NEA assessment of the competitiveness for baseload power generation of nuclear energy against coal- and gas-fired generation under carbon pricing has employed four different methodologies, three of which concentrated on liberalised electricity markets, and has produced a number of results that reflect the perspective of a private investor. The study broadly confirms, albeit in far greater detail and considering a much greater number of variables, the results of the *Projected Costs of Generating Electricity* (IEA/NEA, 2010). And while the Projected Costs study adopted a concept of social resource cost rather than private costs and benefits, the one basic conclusion remains the same: economic competition in electricity markets is today being played out between nuclear energy and gas-fired power generation, with coal-fired power generation not being competitive as soon as even modest carbon pricing is introduced. Whether nuclear energy or natural gas comes out ahead in their competition depends on a number of assumptions, which, while all entirely reasonable, yield very different outcomes.

The only key variable being used for sensitivity analysis in the Projected Costs study was the cost of capital which alternated between real rates of 5% and 10. And unsurprisingly gas-fired power generation was more competitive at a 10% discount rate, while nuclear energy was more competitive at a 5% discount rate. The picture in this study, developed on the basis of daily data from European power markets (including the EU ETS carbon market) over a five-year period, is far more nuanced. Three different methodologies, a profit analysis looking at historic returns over the past five years, an investment analysis projecting the conditions of the past five years over the lifetime of plants and a carbon tax analysis (differentiating the investment analysis for different carbon prices) look at the issue of competitiveness from different angles. They show that the competitiveness of nuclear energy depends on a number of variables which in different configurations determine whether electricity produced from nuclear power or from CCGTs generates higher profits for its investors. They are:

1. *Overnight costs:* the profitability of nuclear energy as the most capital-intensive of the three technologies depends heavily on its overnight costs.[1] This is a characteristic that it shares with other low-carbon technologies such as renewable energies, but the latter are not included in this comparison. The study reflects the importance of capital costs by working with a FOAK case and an industrial maturity case, where the latter's capital cost is two-thirds of the former's.

2. *Financing costs:* since the Projected Costs study nothing has changed on this point. Financing costs have a very large influence on the costs and profitability of nuclear energy. Nevertheless, the study does not concentrate on this well-known point but works (except for one illustrative case) with a standard capital cost of 7% real throughout the study.

---

1.   Capital costs are a function of overnight costs (which include pre-construction or owner's cost, engineering, procurement and construction costs as well as contingency costs) and IDC. The latter depends, of course, on financing costs as discussed under the next point.

3. *Gas prices:* what capital costs are to the competitiveness of nuclear energy, gas prices are to the competitiveness of gas-fired power generation, which spends a full two-thirds of its lifetime costs on fuel. If gas prices are low, gas-fired power generation is very competitive indeed. If they are high, nuclear energy is far ahead. The study reflects this fact by working with a low gas price case and a high gas price case in addition to the base case scenario.

4. *Carbon prices:* low and medium-high carbon prices, up to EUR 50 per tonne of CO_2 increase the competitiveness of nuclear power. However, in contrast to the conclusions of the LCOE methodology employed in the Projected Costs study, high carbon prices do not unequivocally improve the competitiveness of nuclear power in a market environment. As carbon pricing makes coal with its high carbon content the marginal fuel, the revenues of gas increase faster than its cost, with an overall increase in profitability that matches that of nuclear and can surpass it at very high carbon prices.

5. *Profit margins* or "mark-ups" are the difference between the variable costs of the marginal fuel and the electricity price, and are a well-known feature of liberalised electricity markets. They have a very strong influence on the competitiveness of the marginal fuel, either gas or coal, for which they single-handedly determine profits. The level of future profit margins can thus determine the competitiveness between nuclear energy and gas.

6. *Electricity prices:* in a liberalised electricity market, prices are a function of the costs of fossil fuels (natural gas and coal), carbon prices and mark-ups. The higher they are, the better nuclear energy fares, both absolutely and relatively. This is also due to the fact that higher electricity prices go along with higher prices for fossil fuels and carbon.

7. *Carbon capture and storage (CCS):* the standard investment and carbon tax analysis do not assume the existence of pervasive CCS for coal-fired power plants. However, an alternative scenario does and it shows that CCS will remarkably strengthen the relative competitiveness of nuclear energy against gas-fired power generation. The profitability of gas declines significantly once it substitutes for coal as the marginal fuel at high carbon prices.

The particular configuration of these seven variables will determine on the competitive advantage of the different power generation options. The profit analysis showed that during the past five years, nuclear energy has made very substantive profits due to carbon pricing. These profits are far higher than those of coal and gas, even though the latter did not have to pay for their carbon emission permits during the past five years. This will change with the introduction of full auctioning of permits in 2013 in the EU ETS, which will further increase the relative short-term advantage of nuclear power plants. Operating an existing nuclear power plant in Europe today is very profitable.

However, the profit analysis does not take into account investment costs. It is more difficult to summarise the results for the investment and the carbon tax analysis that both take into account the investment costs and compute the costs and benefits over the lifetime of the different plants. Again, a new coal plant is highly unlikely to be a competitive or even a profitable technology option under the price conditions prevailing during the 2005-10 period once it has to pay for its carbon emissions. Concerning the competition between nuclear energy and gas-fired power generation, one needs to be more circumstantiated and refer to the particular configuration of the seven variables presented above. If these seven variables are grouped in three broad categories, investment costs, electricity prices as a function of gas and carbon prices and carbon capture and storage (CCS), then one may summarise the results of the previous chapters in the following manner. *Nuclear energy is competitive with natural gas for baseload power generation, as soon as one of the three categories – investment costs, prices or CCS – acts in its favour. It will dominate the competition as soon as two out of three categories act in its favour.*

Of course, this rough and ready synthesis cannot do justice to the richness of the analysis presented above. Anybody truly interested in the competitiveness of nuclear energy under carbon pricing would be well advised not to bypass the previous chapters. In particular, the previous chapters also develop a number of conceptual issues that have a bearing on the competitiveness between different power generation sources such as the suspension option, the ability to suspend production on days where variable costs fall below prices, or the pass-through of carbon prices into electricity prices.

The competition between nuclear energy and gas-fired power generation remains characterised by the dependence of each technology's profitability on different scenarios. Gas, which is frequently the marginal fuel, makes relatively modest profits in many different scenarios, which limits downside as well as upside risk. The relatively small size of its fixed costs does not oblige it to generate very large profit margins. In addition, the suspension option allows gas to opt out of the market when prices are too low. High electricity prices instead are not necessarily a source for significant additional profits as they frequently result precisely from high gas prices and consequently the high variable costs for gas-fired power plants.

Nuclear energy is in the opposite situation, where its profitability depends very strongly on the level of electricity prices. Its high fixed costs and low and stable marginal costs mean that the profitability of nuclear rises and falls with electricity prices that single-handedly determine its profit margin, the difference between its per-unit revenue and its variable costs. Given that the variable costs of nuclear power are virtually never above electricity prices and it thus has no opportunity to exercise the suspension option, nuclear power will be affected by electricity price changes in a largely passive fashion.

For investors it is thus important to make their own assessment of the probability of different capital costs and price scenarios. If nuclear succeeds in limiting overnight costs and electricity prices in Europe stay high, nuclear is by far the most competitive option. With high overnight costs and low electricity prices, only a very strong logic of portfolio diversification could motivate arguments in its favour. As far as prices are concerned, it is quite likely that European electricity prices will stay high or even increase in the foreseeable future. The progressive exit from both fossil fuels and nuclear in Germany, Europe's biggest market, will inevitably push prices higher, which in conjunction with carbon pricing opens opportunities for nuclear energy in other European countries. Similar dynamics may also assert themselves in the United States, where ambitious greenhouse gas reduction targets also ensure a floor under electricity prices.

A high electricity price scenario is thus likely but by no means assured. In this context, policy makers need to be aware of the fact that the profitability of nuclear energy in liberalised electricity markets depends on specific electricity price scenarios. It is thus not unthinkable that risk-averse private investors may opt for fossil-fuel-fired power generation instead of nuclear *even in cases where nuclear energy would be the least-cost option over the lifetime of the plant.* Liberalised electricity markets with uncertain prices can thus lead to different decisions being taken by risk-averse private investors than by governments with a longer-term view. This especially concerns investments in low-carbon technologies with high fixed costs. The unification and liberalisation of European electricity markets has done much to further the project of European integration and has increased economic welfare through mutualising competitive advantages in baseload and peakload power provision, managerial efficiency and consumer choice in the process. Measures such as long-term contracts for electricity provision could serve to foster the introduction of high fixed cost, low-carbon technologies such as nuclear and large renewables.

An additional aspect of public policy making is provided by the issue of profit margins or mark-ups of electricity prices over the variable costs of the marginal fuel which benefit, in particular, the competitiveness of the last fuel in the merit order. Regardless of whether they are an expression of spontaneous or consciously constructed monopoly power, nuclear energy is favoured by limiting these welfare reducing mark-ups. Market opening and competition in the provision of baseload power provision favour the competitiveness of nuclear energy.

Clearly, also industry has to play its role. With respect to overnight investment costs, for example, the issue is clearly in the court of the main vendors of nuclear power plants, which in Europe will mean inevitably new Generation III+ plants. These plants already have a number of advanced safety features that should satisfy even a substantial tightening of safety requirements in the aftermath of the Fukushima nuclear accident. However, the industry needs to move from a first-of-a-kind scenario to an industrial maturity scenario if nuclear is to stay competitive beyond a scenario of high gas and electricity prices.

In the end, the outcome of the competition between nuclear energy and gas-fired power generation (coal-fired power generation being uncompetitive under carbon pricing), depends on a number of key parameters such as investment costs and prices. The profitability of either nuclear energy or gas-fired power generation, however, cannot be assessed independently of the scenario in which they are situated. Given the realities of the large integrated utilities that dominate the European power market, which need to plan ahead for a broad range of contingencies, the implications are straightforward. Risk minimisation implies that utilities need to diversify their generation sources and need to adopt a portfolio approach. Any utility would thus be advantaged by adopting a portfolio approach. Such diversification would not only limit financial investor risk, but also a number of non-financial risks (climate change, security of supply, accidents). Portfolio approaches and the integration of non-financial risks will thus both be important topics for future research at the NEA and in the wider energy community.

Bibliography

Brealey, R., F. Allen and S. Myers (2006), *Principles of Corporate Finance*, Irwin: McGraw-Hill, United States.

Burtraw, D. and K. Palmer (2007), "Compensation Rules for Climate Policy in the Electricity Sector", *Discussion Paper 07-41*, Resources for the Future (RFF), Washington DC, United States.

Cambini, C. and L. Rondi (2010), "Incentive Regulation and Investment: Evidence from European Energy Utilities", *Journal of Regulatory Economics*, 38 (1), pp. 1-26.

CRE (2010), "Délibération de la Commission de régulation de l'énergie du 28 octobre 2010 portant proposition de modification des tarifs d'utilisation des réseaux de transport de gaz naturel", Commission de régulation d'énergie, France, www.cre.fr/fr/documents/deliberations.

Dixit, K.A. and R.S. Pindyck (1994), *Investing under Uncertainty*, Princeton University Press, Camden (NJ), United States.

European Commission (EC) (2010), *Emissions Trading (EU ETS)*, http://ec.europa.eu/environment/climat/emission.

Ellerman, D., F. Convery and C. de Perthuis (2010), *Carbon Pricing: The European Union Emissions Trading Scheme*, Cambridge University Press, Cambridge, England.

Geman, H. (2005), "Spot and Forward Electricity Markets", *Commodities and Commodity Derivatives: Modelling and Pricing for Agriculturals, Metals and Energy*, Wiley, Chichester, England, pp. 251-282.

Green, R. (2008), "Carbon Tax or Carbon Permits: The Impact on Generators' Risks", *Energy Journal*, 29(3), pp. 67-90.

Hicks, J. (1932), *The Theory of Wages*, Macmillan, London, England.

IEA (2007), *Climate Policy Uncertainty and Investment Risk*, International Energy Agency, OECD, Paris, France.

IEA (2009), *World Energy Outlook 2009*, International Energy Agency, OECD, Paris, France.

IEA (2010a), *CO_2 Emissions from Fuel Combustion*, International Energy Agency, OECD, Paris, France.

IEA (2010b), *Reviewing Existing and Proposed Emissions Trading Schemes*, International Energy Agency, OECD, Paris, France.

IEA/NEA (2010), *Projected Costs of Generating Electricity: 2010 Edition*, OECD, Paris, France.

IPCC (2007), "Fourth Assessment Report: Climate Change 2007", Working Group III Report Mitigation of Climate Change, Intergovernmental Panel on Climate Change, CUP, Cambridge, England, www.ipcc.ch/publications_and_data/ar4/wg3/en/contents.html.

Joskow, P. (2006), "Competitive Electricity Markets and Investment in New Generating Capacity", MIT Working Paper, MIT, United States, http://econ-www.mit.edu/files/1190.

Matthes, F.C. (2008), "Windfall Profits of German Electricity Producers in the Second Phase of the EU Emissions Trading Scheme (2008-2012)", Briefing Paper for World Wide Fund for Nature Germany, Oko-Institut e.V., Berlin, Germany at www.oeko.de/oekodoc/760/2008-222-en.pdf.

Keppler, J.H. and M. Cruciani (2010), "Rents in the European Power Sector Due to Carbon Trading", *Energy Policy*, 38 (8), pp. 4280-4290.

Keppler, J.H. and M. Mansanet-Bataller (2010), "Causalities between CO_2, Electricity, and other Energy Variables during Phase I and Phase II of the EU ETS", *Energy Policy*, 38 (7), pp. 3329-3341.

Pozzi, C. (2007), "The Relationship Between Spot and Forward Prices in Electricity Markets", Chapter 9, *The Econometrics of Energy Systems*, J. H. Keppler, R. Bourbonnais and J. Girod Eds, Aldershot: Palgrave Macmillan, England.

Rogner, H.-H. and A. McDonald (2008), "Long-term Performance Targets for Nuclear Energy (Part 2): Markets and Learning Rates", *International Journal of Global Energy Issues*, 30 (1-4), pp. 77-101.

Roques, F.A., D.M. Newbery and W.J. Nuttal (2008), "Fuel Mix Diversification Incentives in Liberalized Electricity Markets: A Mean-Variance Portfolio Theory Approach", *Energy Economics*, 30, pp. 1831-1849.

Roques, F.A., W.J. Nuttal, D.M. Newbery, S. Connors and R. de Neufville (2006a), "Nuclear Power: A Hedge against Uncertain Gas and Carbon Prices?", *The Energy Journal*, 27 (4), pp. 1-24.

Roques, F.A., W.J. Nuttal and D.M. Newbery (2006b), "Using Probabilistic Analysis to Value Power Generation Investments under Uncertainty", *Cambridge Working Papers in Economics*, p. 650.

Rothwell, G. (2006), "A Real Options Approach to Evaluating New Nuclear Power Plants", *Energy Journal*, 27 (1), pp. 37-53.

World Bank (2010), *State and Trends of the Carbon Market 2010*, World Bank, Washington DC, United States, http://siteresources.worldbank.org/INTCARBONFINANCE/Resources/State_and_Trends_of_the_Carbon_Market_2010_low_res.pdf.

Yang, M. and W. Blyth (2007), "Modeling Investment Risks and Uncertainties with Real Options Approach", IEA Working Paper, International Energy Agency, OECD, Paris, France.

Annex I

Acronyms

ARA	Amsterdam-Rotterdam-Antwerp
CCGTs	Combined cycle gas turbines
CCS	Carbon capture and storage
CDM	Clean development mechanism
CER	Certified emission reductions
CH_4	Methane
CO_2	Carbon dioxide
EPC	Engineering, procurement and construction
EU ETS	European Emissions Trading System
EUAs	EU Allowances
FOAK	First-of-a-kind
IDC	Interest during construction
IRR	Internal rate of return
LCOE	Levelised cost of electricity
LR	Learning rate
MIRR	Modified internal rates of return
NEA	Nuclear Energy Agency
NPV	Net present value
OC	Overnight investment cost
OECD	Organisation for Economic Co-operation and Development
PI	Profitability index
RR	Reinvestment rate
SCR	Selective catalytic reduction
TICAP	Total installed capacity
US RGGI	US Regional Greenhouse Gas Initiative (Northwest US)
WACC	Weighted average cost of capital

Annex II
List of experts

BELGIUM

Gilbert CORNELISSEN   SA Synatom

Chantal CORTVRIENDT   SPF Economie

CANADA

Stella LAM   Atomic Energy of Canada Limited (AECL)

Lilian TARNAWSKY   Atomic Energy of Canada Limited (AECL)

CZECH REPUBLIC

Lubor ŽEŽULA   Nuclear Research Institute Řež plc

FRANCE

Philippe LEBRETON   EDF-CIST

Frédéric LEGEE   CEA Saclay

François PERFEZOU   Ministère de l'Écologie, de l'Énergie, du Développement

Stephane Rouhier   General Directorate for Energy and Climate Change Carbon Markets Division (MEEDDM)

GERMANY

Alfred VOSS (Chairman)   Universität Stuttgart, Institut für Energiewirtschaft und rationelle Energieanwendung

Johannes KERNER   Bundesministerium für Wirtschaft und Technologie

HUNGARY

György WOLF   Paks NPP

Karoly GERSE   Hungarian Power Companies Ltd.

ITALY

Fortunato VETTRAINO   Ente per le Nuove Tecnologie l'Energia e l'Ambiente ENEA

JAPAN

Dr. Kazuaki MATSUI   The Institute of Applied Energy (IAE)

Mr. Koji NAGANO   Central Research Institute of Electric Power Industry (CRIEPI)

KOREA (REPUBLIC)

Mr. Hun BAEK	Korea Hydro & Nuclear Power Co., Ltd (KHNP)
Dr. Jin-Ko HO	Korea Hydro & Nuclear Power Co., Ltd (KHNP)
Kun Jai LEE	Korea Advanced Institute of Science and Technology (KAIST)
Mankin LEE	Korea Atomic Energy Research Institute (KAERI)
Kee Hwan MOON	Korea Atomic Energy Research Institute (KAERI)

NETHERLANDS

Gert van UITERT — Ministry of Economic Affairs

POLAND

Olgierd Skonieczny	Nuclear Power S.A.
Andrzej STRUPCZEWSKI	Institute of Atomic Energy POLATOM

SPAIN

Antonio GONZÁLEZ JIMÉNEZ — Spanish Nuclear Industry Forum

SWITZERLAND

Michel DELANNAY	Kernkraftwerk Gösgen-Däniken AG
Roger J. LUNDMARK	Swissnuclear

UNITED STATES

Matthew P. CROZAT, (Chairman) — Office of Nuclear Energy, US DoE

European Commission (EC)

Christian KIRCHSTEIGER — DG TREN-H2

International Atomic Energy Agency (IAEA)

Nadira BARKATULLAH — Department of Nuclear Energy

International Energy Agency (IEA)

Robert Arnot	Energy Analyst, Electricity Energy Markets and Decarbonisation
Shinichiro KADONO	Consultant, EDD

OECD Nuclear Energy Agency (NEA)

Ron Cameron	Head, Nuclear Development Division (NDD)
Jan Horst KEPPLER	Principal Economist, NDD
Claudio MARCANTONINI	Consultant, NDD